GEOLOGY EXPLAINED IN
THE SEVERN VALE
AND COTSWOLDS

Dedicated to
L. RICHARDSON, F.R.S.E., F.G.S.

who, in the early part of this century, did so much pioneer work on the geology of the Cheltenham area and whose book *The Geology of Cheltenham* 1904 remains a classic on this locality.

GEOLOGY EXPLAINED IN THE SEVERN VALE AND COTSWOLDS

by

WILLIAM DREGHORN, *B.Sc., F.R.G.S.*

*Illustrations by
the Author*

DAVID & CHARLES : NEWTON ABBOT
1967

Acknowledgments

I would like to express my grateful thanks to the two people who have assisted in the writing of this book:

To the eminent geologist, Dr D. V. Ager, who so generously spared time to read and criticise the final script in spite of many pressing commitments in more exalted fields of study.

To Mr C. E. Leese, B.Sc., retired headmaster, past president of Royal Geological Society of Cornwall and literary critic of *The Western Morning News*, who read and criticised the script while the book was taking shape.

W.D.

Printed in Great Britain
by W. J. Holman Limited Dawlish
for David & Charles (Publishers) Limited
Newton Abbot Devon

Contents

Introduction

It is an old teaching maxim that 'if you want to make a subject live, you must make it local'. The purpose of this book then, is to make geology a living subject by making it local.

Students of geology will find it useful but as it has been written for all who are interested in scenery, the maps, diagrams and illustrations have been simplified so that they are also intelligible to the layman. Again, the geological expert usually studies rocks by looking at exposures in quarries, railway-cuttings and boreholes, as well as by looking at scenery, whereas this book attempts to introduce geology to the layman in terms of scenery, and explains the relationship of rocks to hills, valleys and plains, and the way in which rock structures have influenced the evolution of different types of scenery.

Only a few maps have been included among the illustrations as I have assumed that readers who are geologists will purchase the relevant one-inch-to-the-mile Ordnance Survey geological maps, the Moreton-in-Marsh and the Cirencester sheets. Instead of maps, I have, whenever possible, used 'block' diagrams, which non-geologists often find easier to interpret than geological maps.

When drawing a block diagram the artist selects a 'block' of landscape and cuts through it with an imaginary knife to expose the rocks of the strata underlying the scenery, at the same time putting in identifying details of the scenery above. When I used this method to illustrate a series of newspaper articles on Cotswold villages written by my wife, many villagers found that these block diagrams had helped them to understand their environment in a way no map had ever been able to explain it.

One legitimate criticism which could be levelled at this book is that comparatively little attention has been paid to the characteristic fossils of the areas covered—particularly as the geology books written by Gloucestershire's nineteenth-century geologists mainly consist of lists of fossils. However, listing fossils has not been 'the object of the exercise' in this case, and the reader who requires more information on this subject is recommended to buy the British Museum booklets on fossils.

The choice of the area covered by this book is explained by the fact that Gloucestershire and its adjoining counties are exceedingly fortunate in having a remarkably wide variety of rocks all within easy access. The geology student who takes his degree at Bristol University can see almost the entire range of rocks in the British Isles within fifty miles of Bristol whereas, in areas like Australia or New Zealand, the student must travel hundreds of miles over the same rocks before a change can be observed.

Within fifty miles of Gloucester, Cheltenham or Bristol there can be seen the following wide range of rocks:

1. The Pre-Cambrian rocks of the Malverns which are some of the oldest rocks in the British Isles; they form part of the original framework of the British Isles constructed by earth movements many millions of years ago.
2. The Silurian rocks around May Hill, the Malverns and Ledbury.
3. The Carboniferous rocks of the Lower Severn and the Avon Gorge.
4. The Old Red Sandstone rocks of the Forest of Dean.
5. The Coal Measures of the Forest of Dean.
6. The Jurassic rocks of the Cotswolds.
7. The Triassic rocks of the South Midlands.
8. The Cretaceous rocks (the chalk hills of Wiltshire).
9. The finest exposures in the whole of the British Isles of the Rhaetic rocks. These are in the Lower Severn area at Aust and Westbury-on-Severn and contain the famous 'bone bed', rich in the fossil remains of fishes and reptiles.

PHYSICAL FEATURES OF THE AREA

The main physical features of the area covered by this book are shown in the sketch at Map 1 and include the Severn Vale bordered on the east by the strong escarpment of the Cotswolds and, to the west, the hilly regions of the Welsh borderlands. These two main features guide the River Severn in its south-westerly course to the Bristol Channel. To keep the sketch map clear and simple, the numerous valleys of the Cotswolds are not shown, but east of the main escarpment the drainage is to the Thames.

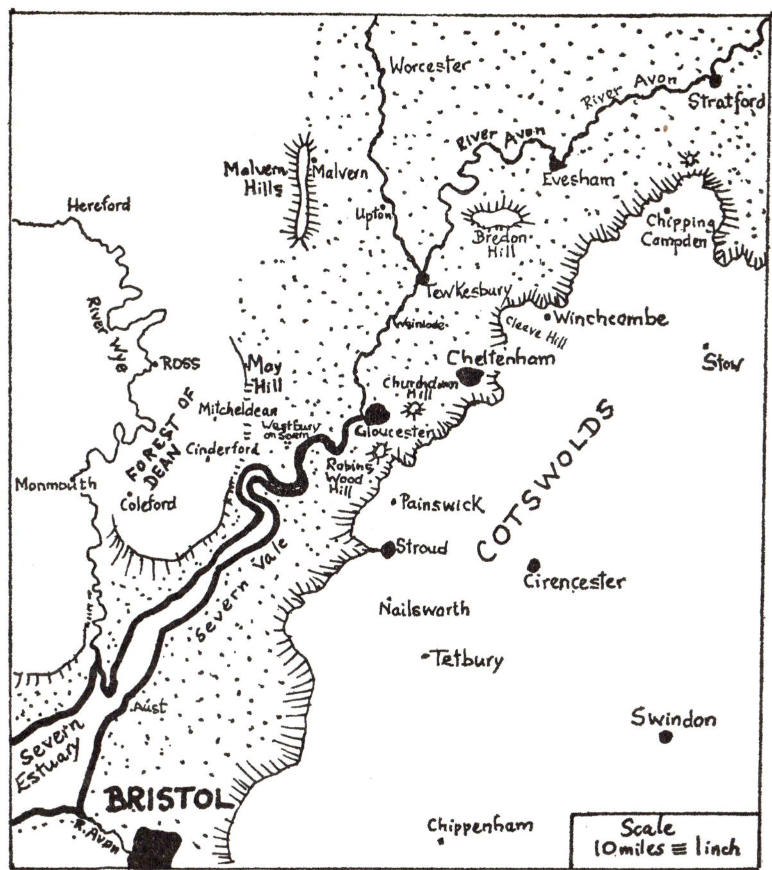

MAP 1 Sketch map of the Cotswolds & Severn Vale

The corresponding geological map (Map 2) shows the control exercised by geology in forming those relief features. In general, the high ground of the Welsh borderlands is formed of older or Palaeozoic rocks which, being harder and more resistant to erosion, stand out as higher ground. The low-lying areas of the Severn Vale, on the other hand, consist of the softer clays of the Trias and Lower Lias, and clearly demonstrate how rivers tend to erode their valleys in the softer rocks. Harder rocks of the Oolitic limestones are responsible for the high ground of the Cotswolds and these form a plateau, or 'dip-slope', which falls away gently south-eastwards down to the plain of Oxford.

By comparing the two maps it will be seen that the Cotswold escarpment has been eroded back to leave remnants of the former scarp, including such conspicuous hills as Churchdown and Robin's Wood, rising castle-like out of the Severn Vale. These remnants,

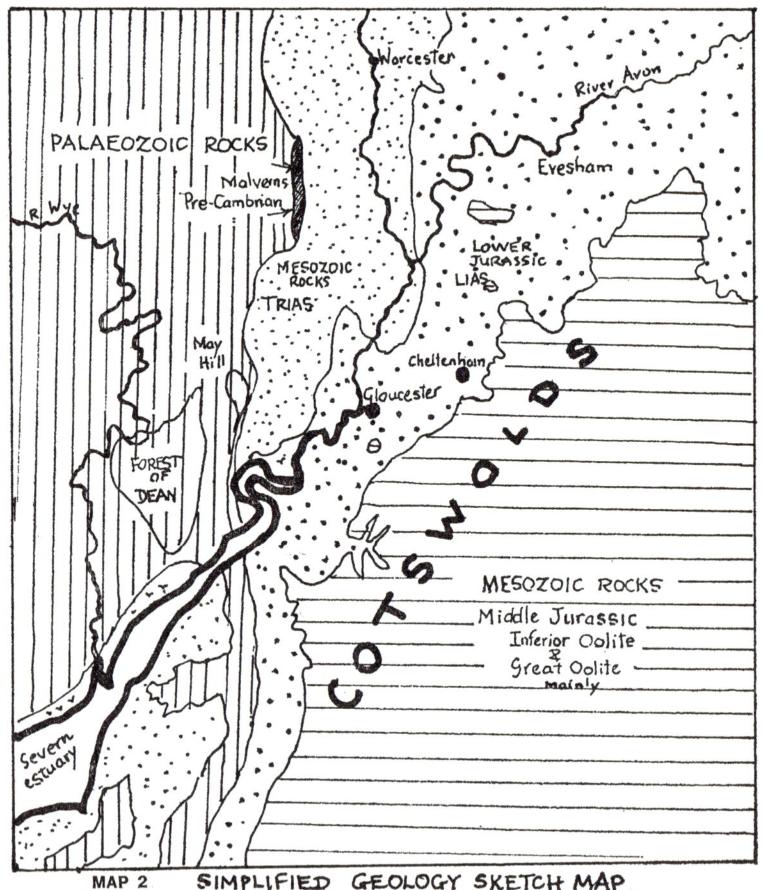

MAP 2   SIMPLIFIED GEOLOGY SKETCH MAP
Older Palaeozoic rocks on the west.
Younger Mesozoic rocks on the east
Recent alluvium and glacial deposits (not shown) in the Severn vale.

known as 'outliers', are the subject of a later chapter.

The Malvern range, too, is a very marked feature and, composed of ancient crystalline rocks rising suddenly from the plains of the Trias, is really part of the backbone of England. The Australian term for such a range would be 'jump-up' and, geologically, these hills were pushed up millions of years ago.

Looking at the physical map, one may ask where is the best scenery to be found? The answer depends upon what you fancy, for there is certainly no lack of variety. The highest parts of the Cotswolds are on Cleeve Hill, 1,083 ft above sea level, and these are fine scenic areas; for more rugged scenery the Malverns are the answer and, in parts, resemble the highland zone of the Welsh interior. Craggy, limestone country is to be found in the Chepstow and Wye valley, where the massive Carboniferous limestones are deeply cleft by the incised meanders of the River Wye. There are some delightful spots, too, along the banks of the Severn, where the red rocks of the Trias (the Keuper Marl) form cliffs at Wainlode, Westbury-on-Severn and at Aust, further evidence that the variety of scenery in this region is mainly due to the great variations in the type of rock.

A geological map is rather like Jacob's Coat, a thing of 'many colours', with a different colour for each different period of rock formation—and the prettiest maps with the greatest variety of colours are undoubtedly those of the areas in and around Gloucestershire.

CHAPTER 1

The Jurassic System of Gloucestershire

Most of the rocks in the eastern half of Gloucestershire belong to the Jurassic system—so named after the rocks in the Jura Mountains in France which are of the same age.

There are several main aspects of any rock system which have to be studied in a geological assessment, and these are:

STRATIGRAPHY, or descriptions of the various bands or strata of rock and the relationship between those bands;

PETROGRAPHY, which is the study of the nature of the materials in the bands; and

PALAEONTOLOGY, the study of the fossils found in the strata.

Palaeontology serves stratigraphy in that a bed of rock (or stratum) can be identified by the fossils it contains. Then, by studying the fossils as complete assemblages and not merely as individual fossils, the geologist endeavours to arrive at a conclusion about the life of the time when they were deposited.

This principle of identifying strata by means of fossils was first pointed out by William Smith, the son of an Oxfordshire blacksmith who became a surveyor and civil engineer during the great days of canal-building at the end of the eighteenth century. He was engaged in survey work for the building of the Somerset Coal Canal from 1792 to 1795 and throughout that period he made a profound study of the Jurassic rocks around Bath, having been from earliest boyhood an observer and collector of Jurassic fossils—even playing marbles with them! It was he who was the first to recognise that it is 'a general law that the same strata are found always in the same order of superimposition and contain the same peculiar fossils' —thereby laying the foundations for the new science of geology.

The greatness of William Smith did not rest upon book-learning, but came from his powers of accurate observation, originality of thought and his constructive imagination. He was less concerned with devising appropriate names for his fossils than he was with perceiving their place in the ordered sequence of the rocks which he and his fellow-engineers were uncovering. He was a practical man whose learning was alive and direct and, above all, he was an original creative thinker.

The terms which William Smith chose to describe the layers of rock he uncovered were merely words borrowed from the language of the common workers of the day, workers in agriculture and the quarrying and building industries. In fact, in 1831, when the savants of the Geological Society presented him with a medal, his phraseology was described as 'those arbitrary and somewhat uncouth terms which we derive from him as our master'. Nevertheless, most of those terms are still in use today.

The most important point to remember about stratigraphy is that when various strata are lying one above the other, the older rocks are below and the younger are on top, except when it can be seen that the whole mass may have been turned upside down by some earth movement which has tilted, folded and dislocated the beds of rock.

CLAYS, SANDS AND LIMESTONES

The Cotswold hills are composed of limestone and the low-lying plains of the Severn Vale are of clay. The foothills are sandy rocks. This is a neat order of clays, sands and limestones—and is repeated many times throughout the series of Jurassic rocks, which were formed when the area they now cover was under the sea. This recurring order of clay, sand, limestone is known as a 'rhythmic succession', and these different substances tell geologists what kind of sea was once there. Clays, for instance, denote a muddy sea, rather deep, whereas sands indicate a shallower sea with incoming rivers bringing sand from the surrounding hills. Limestones represent clear seas, shallow like the Bahamas on a continental shelf.

Each stratum, whether clay, sand or limestone, has its own assemblage of fossils which indicates the conditions existing at the time of deposition. For example, most of the ammonites liked to live in muddy seas. So did the marine reptiles, because plenty of food was thus available. On the other hand, corals, sea lilies and animals which liked to live and feed in clear seas are most plentiful in the limestones of the Cotswolds.

Figures 1 & 2 overleaf show how the rocks of the Cotswolds 'strike' down the country from north to south and at the same time dip gently eastward, so that the upturned edges face Wales. And remembering that the strike of rocks is always at right angles to the

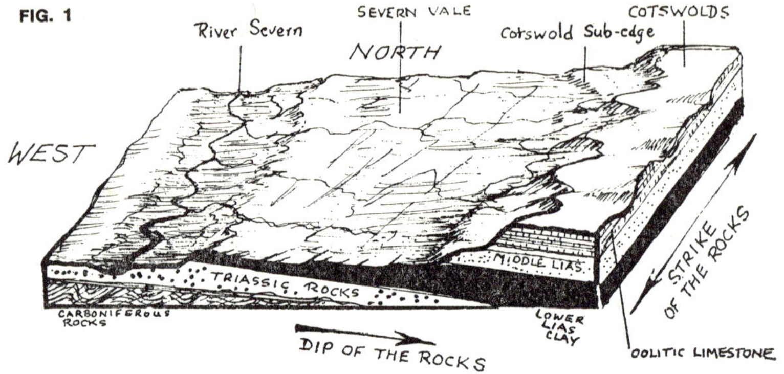

GEOLOGY OF THE REGION — VERY SIMPLIFIED

The course of the River Severn is generally controlled by the STRIKE OF THE ROCKS. The rivers flowing across the plain to join the Severn are flowing against the DIP of the rocks and consequently sometimes meet obstructions of upturned beds. This caused "rapids" of a very minor nature but good enough to establish mills.

dip or inclination of the strata, they also show how this strike controls the direction in which the Severn flows.

It will be noticed that rocks older than the Jurassic appear on the surface to the west of the region. These are called Triassic rocks and, if traced northwards, they will be seen to form the Plain of Cheshire. It can also be seen that wherever they appear the clay plains form a great corridor route for rivers, railways and motorways from the Midlands to Bristol. And, since transport facilities are determining factors in the location of industry today, this is one good reason why industries and people from the overcrowded Midlands are now spilling down this corridor.

THE LOWER LIAS CLAY

The clay forming the plains has thin bands of limestone which used to be quarried by layers. Eighteenth-century quarry workers called this type of rock the 'Lias' and William Smith borrowed the term. The Lower Lias clays form the plains of the Jurassic system and the lower shelf of sands and sandy limestones are called the Middle Lias. The Cotswolds are the Oolitic limestones of the Middle Jurassic system, the nature of which is explained later on in this book.

The Lower Lias clay is a blue clay which was laid down in a muddy sea some 170 to 180 million years ago. It is about 600 ft thick in the Cheltenham/Gloucester area and reaches a maximum of 960 ft near Evesham. Towards Bristol it thins out to 100 to 200 ft. This is because the muddy, shallow sea was constantly changing its shore-lines and, at times, silting up to swamps and deltas as sediment from the constant erosion of the high mountains in the surrounding lands of the Mendips and Wales was deposited in it. Sometimes, too, the rivers carried into the sea limey muds from the products of erosion of limestone mountains like those of the Mendips which, today, are merely the remains of much more extensive mountains. This is why we often find bands of hard limestone in the Lower Lias clay and why, when motoring over the plains from Cheltenham towards Bristol, we see gentle undula-

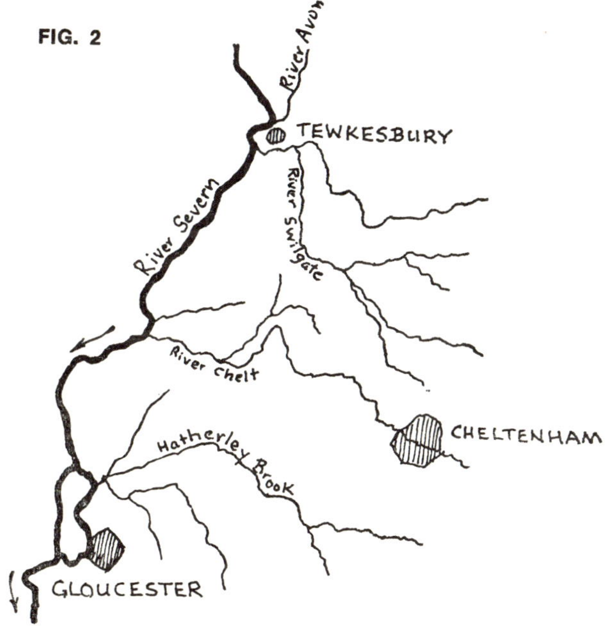

SKETCH MAP TO SHOW HOW THE
PATTERN OF RIVERS AND STREAMS
REFLECTS THE UNDERLYING GEOLOGY
COMPARE THIS WITH THE O.S. geological map.

tions caused by these hard bands. It is not a flat plain and is known to most people as the Severn Vale.

These limestone bands increase in number towards the south and are best seen in the clay cliffs near Frampton-on-Severn (Figure 3). This area is called Fretherne, and here one can walk along the fore-shore of the River Severn and obtain many fossils from the lime-stone bands. It is very rare, however, to find exposures of the Lower Lias clays, for whenever a cutting is made, as for a brickyard or a railway line, the exposure is soon covered with grass.

The Rivers Chelt and Swilgate, and a few other left-bank tributaries of the Severn, all show an angular pattern on the map and this could be related to the underlying structures formed by the hard bands in the Lower Lias clay. There is also the possibility of an old drainage system having been disorganized by glacial action.

In some cases the small stream tumbles over a hard band of clayey limestone (argillaceous limestone) and here the rapids formed by the break of slope led to the establishment in the Middle Ages of important mills, among them Slate Mill at Boddington, Barrett's Mill in Cheltenham, Sandford Mill in Charlton Park, and the mills at Brockworth and Stoke Orchard.

In some cases the mill sites are at breaks of slope caused by a special feature of river erosion termed by geomorphologists 'nick points'—where a sudden drop in the river profile results from changes in the base level of the master river, in this case the Severn.

THE AGE OF THE ICHTHYOSAURUS

The former muddy seas of the Lower Lias clay lasted for about ten million years and in them, 180 million years ago, lived many kinds of ammonites and reptiles, including many types of oysters, some being quite curiously curved. But it is rare indeed ever to find a large animal preserved in its entirety in the strata because when it died and fell to the bottom of the sea it was almost invariably devoured by scavengers. Moreover, in large organised animals the skeleton is held together by bands of muscle and when these decompose, the articulated bones are released and distributed by the currents on the floor of a shallow sea.

The most common animal at that time was a marine reptile called Ichthyosaurus, which often grew up to twenty-five or thirty feet in length. The plaster cast of one found in the Bristol region is now in

FIG. 3

the Cheltenham Museum, and there is the actual fossil of another in the church porch in the village of Tredington—where it is being worn away by the steps of the faithful!

The Ichthyosaurus lived on fish, sea-urchins, and creatures like the sea squids, of which the only hard part left behind as a fossil is its guard, called 'belemnite', which can often be found in the clays. Like other mammals, such as our contemporary whales and dolphins, which returned to the sea in the course of evolution, the Ichthyosaurus was a viviparous reptile—that is, it produced living young in an advanced stage of development—which gave up the land with its constantly shifting shore-lines, and took to an aquatic life.

B

CHAPTER 2

Wainlode Cliff

Cheltenham prides itself on being the tourist centre for the Cotswolds and the Severn Vale, and in order to simplify direction finding it will be assumed that the reader is either living or staying in the Cheltenham/Gloucester area.

Thus, to get to Wainlode Cliff from Cheltenham, one takes the Tewkesbury road and drives across the clay plains of the Lower Lias. There is a sudden rise at Coombe Hill which is caused by the tilted nature of the rocks, hard bands of limestone forming part of a rock division known as the Rhaetic.

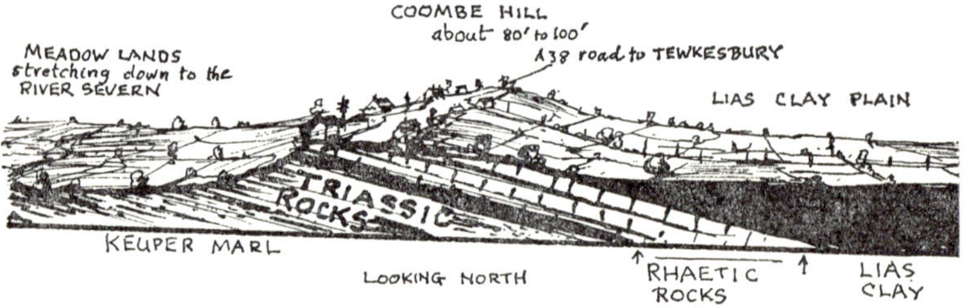

FIG. 4 GEOLOGICAL SECTION ACROSS COOMBE HILL

These Rhaetic beds form a minor escarpment which provided a convenient causeway for building that part of the A38 main road which runs between Gloucester and Tewkesbury. Driving along the top of this scarp, you will enjoy some remarkable views—on one side you look down across the fields towards the Severn and on the other there are extensive views across the plains towards the Cotswolds and Bredon Hill.

To reach Wainlode Cliff it is best to take the turning to Haw Bridge, branching off to the left before reaching the bridge itself. Here the river swings alongside fine red cliffs by the Red Lion Inn, in front of which green meadows stretch down to the river's edge. This is Gloucester's 'Lido', which becomes quite crowded on sunny

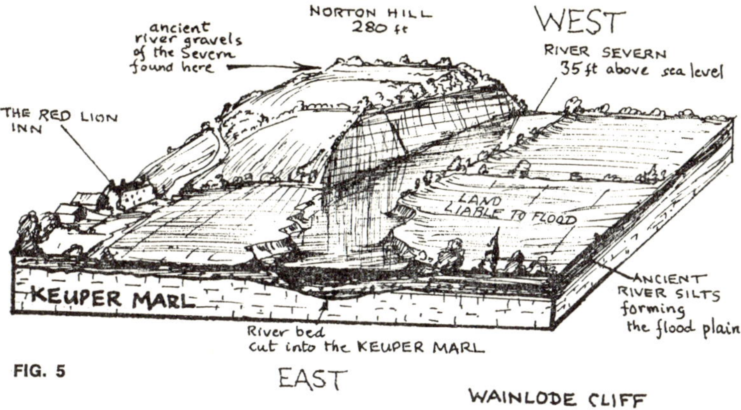

FIG. 5

weekends, and is no place to go to after heavy winter rains. For then the surrounding fields become flooded, parts of the road are impassable and, when the flood waters subside, thin coats of mud are left behind which, in course of time, accumulate to form plains of mud compacted into layers.

Just in front of the Red Lion Inn is a good place for 'rock spotting', and if you scramble down to the water's edge over the ten-foot high cliffs you will see all the layers of sediment laid down from ancient floods. These layers, some of which contain shells of the freshwater mussel, Unio, represent the passage of several thousand years, but in relation to 'geological' time you will be looking at deposits which are very recent indeed.

There is considerable erosion of the river banks here, caused by the speed of the current and tidal effects. And as the River Severn is tidal right up to Tewkesbury, the wash created by passing boats supplements and accelerates the normal erosion process.

In some layers of the alluvium—the name for a recent deposit of river mud—small pebbles of coal can be dug out, and as their origin is obscure they are often the subject of profound discussions in the bar of the Red Lion. 'Old stagers' will tell you that over 100 years ago barges came up the river carrying coal and that some of this could have dropped into the river—although how coal pebbles could be lodged five feet down is a mystery, unless the places where this occurs is 'made up' ground. To the geologist, the thought occurs that the coal pebbles might have been transported by the river in

quite ancient times, possibly having been brought down from the
coalfields of the Midlands.

THE RED CLIFFS AT WAINLODE

Another topic in the inn is 'Why are the cliffs red?' and here the
geologist is on firmer ground. The rocks are red because they were
laid down in a vast desert some 190 million years ago in the period
known as the Triassic. The rocks of any desert always show bright
colours—reds, browns or yellows—because there is not enough
rainfall to carry all the colour-bearing minerals away in solution.
The scene in the Triassic desert would have been something like
the sketch below.

It was a vast plain with salt lakes evaporating under a fierce sun.
Here and there mountain ranges appeared as 'islands' (inselbergs),
the Malverns in those days appearing as a much larger north/south
range. When the lakes to the north dried up, vast deposits of salt
were left behind and these are used by man today in the Triassic
rocks of the Cheshire Plain.

The Wainlode red cliffs are deposits of very fine lake muds, rich

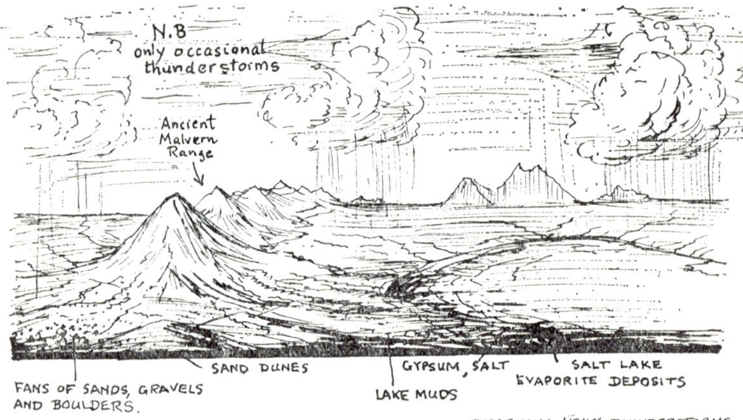

FIG. 6   A TRIASSIC LANDSCAPE

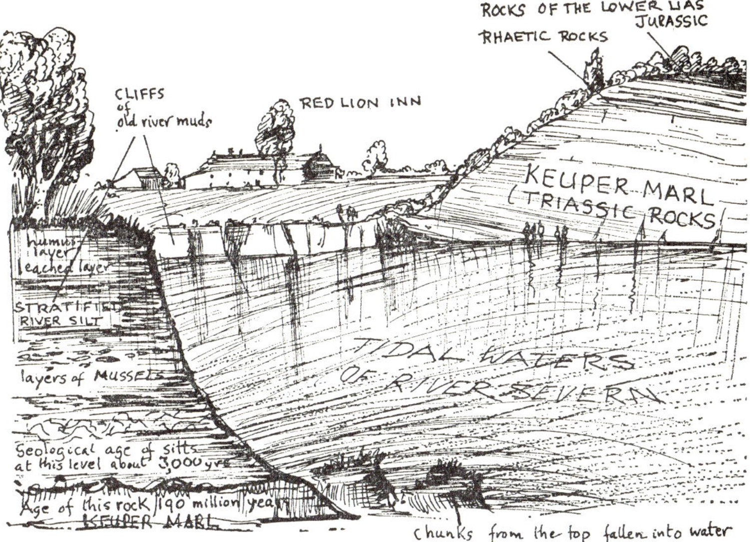

FIG. 7   WAINLODE CLIFF

in iron oxides—and of such fineness that it weathers in a most peculiar way, and crushing it becomes a game of endless fascination for visiting children. Playing along the cliffs, they pick up pieces of the rock and can watch it crumbling from cube shapes to cubes with rounded corners and then right down to small, marble-like spheres which finally shatter to fragments. This process is called 'cuboidal weathering'.

Visitors to Wainlode Cliff can actually hear and see erosion taking place, particularly on summer evenings after a fine day when the sun has been shining on the cliffs. As the temperature drops in the evening there is expansion and contraction of the cliff surface and little pieces of the rock come tumbling down. Erosion of both cliffs and banks is very rapid here owing to the extreme softness of the rock, and the landlord of the Red Lion estimates the loss of the banks of the river to be as much as one yard each year.

The Germanic-sounding word, 'Keuper', which appears in the sketches in this chapter, is derived from the name given to one of the divisions of the Triassic deposits in Germany, and the term

'Triassic' itself comes from Germany where these deposits occur in *three* main divisions: Bunter Sandstone, Muschelkalk (shell limestone) and Keuper.

The term 'marl', however, is really a misnomer perpetuated from earlier usage, because this particular material is, in fact, a very fine-grained, wind-borne dust or lake mud with hardly any stratification, i.e. almost structureless. Again, marls are calcareous whereas this Keuper Marl is not, and breaks with a starchy fracture which may be related to the mineral dolomite in the rock—dolomite is calcium and magnesium carbonate. And this may possibly account for that curious cuboidal weathering which, as yet, is one of the unsolved mysteries of the geologist's world.

THE TEA GREEN MARLS

The Red Marls are about seventy-five feet thick but above them the colour markedly changes to green. These rocks are known as the Tea Green Marls and they are about twenty-three feet thick at Wainlode. To explain the cause of this change in colour we have to turn to chemistry, for the colours in the Wainlode Cliff rocks are due to the presence of minerals. The chief colour-forming mineral in this case is oxide of iron, of which there are several kinds.

Red to brown minerals are fer*ric* oxides, and green minerals are fer*rous* salts, and where you have oxidising conditions you tend to get the red-to-brown minerals, i.e. the fer*ric* condition. If, however, as often occurs in stagnant waters, oxygen is taken *away* from the minerals by bacteria, you then have what is known as 'reducing' conditions and when the oxygen is reduced the materials go back to fer*rous* compounds, which are green. And it is the writer's belief that the green colour in the Tea Green Marls is due to the presence of ferrous hydroxide $(Fe(OH)_2)$. The necessary reducing conditions must have occurred in the swampy lagoons as the Rhaetic seas came in, and these are referred to again in a later chapter on Garden Cliff, Westbury-on-Severn.

To put this interpretation in another way: red rocks of the desert containing ferric minerals were deposited in lakes under clear water and good oxidising conditions. Later, the lakes became stagnant with vegetation because the climate was changing to a more humid one and *the sea was now beginning to invade the land*. Bacteria became active in the rotting vegetation around the lagoons and the oxygen was taken from the red ferric minerals, thereby changing them to the green ferrous condition.

There are no fossils in these rocks, but the writer has found leaf impressions of the primitive plant *Naiadata* in the red rocks. A small, bivalved crustacean called *Euestheria minuta* also lived in the lakes and can be found in the upper part of the cliff.

THE RHAETIC ROCKS

The sea came in slowly, gradually engulfing areas where conditions were similar to those found in the Dead Sea today. Lagoons with vegetation were formed and life began to return in abundance.

This metamorphosis can be more easily understood by reference to the different rock systems depicted in Figure 8. Notice the sudden

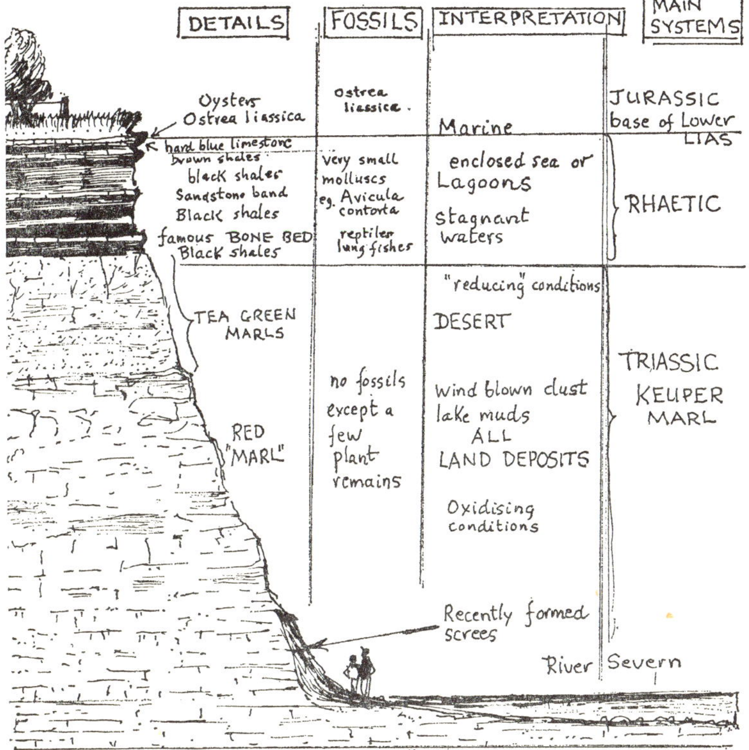

FIG. 8   SIMPLIFIED SECTION OF WAINLODE CLIFF

change to black rocks in the Rhaetic system. These rocks are thin, paper-like shales, ancient coastal muds, black with the mineral ferrous sulphide. In this region, between Aust Cliff and Wainlode Cliff, were congregated crowds of fishes, lung-fishes and marine reptiles. Under these crowded conditions the carnivores ate up the herbivores and their remains were compressed into the thin band of rock known as the 'bone bed'. Because deposition in the lagoon was very slow, the bone bed is very thin here, and as it can be seen to better advantage at Westbury-on-Severn (Garden Cliff) it will be more fully described in the next chapter.

Further up the cliff in the Rhaetic rocks will be found bands of sandstone with marine fossils, then more grey shales and, finally, the blue limestone band which is the 'evidence' of the Lias sea of the Jurassic system. This band has oyster fossils called *Ostrea liassica* and is the base of the Jurassic formation known as the Lower Lias. Just below, one can pick out the mineral gypsum, in the form of very small selenite crystals. This has been formed by percolating water dissolving out iron sulphide which, interacting with calcite from calcareous minerals, produced calcium sulphate, which is gypsum.

Before you leave Wainlode Cliff look out for 'fools' gold', in the form of brassy-yellow crystals of pyrites (iron sulphide, FeS_2). They occur in clusters in some of the Rhaetic shales and some are as big as Oxo cubes. They tumble down the cliff and can often be picked up in the river simply by dredging near the water's edge with one's hand. From time immemorial, men have been fooled into believing that this was truly gold whereas, in fact, these pyritised shales are, again, the result of stagnant conditions, poor in oxygen, their different colour in this case being due to a different combination of chemicals.

CHAPTER 3

Westbury-on-Severn

Westbury-on-Severn is about six miles south-west of Gloucester on the right bank of the Severn. It is reached by taking the A48 main road after crossing the river at Gloucester, and there is a footpath across the fields which leads past Westbury's charming church.

The cliffs by the river are the same rocks in age and type as those at Wainlode Cliff—the red Triassic rocks and Keuper Marl—but the rocks of the 'Garden Cliff' at Westbury are even more striking in appearance. This is no place, like Wainlode Cliff, for bucolic merriment and cheerful curiosity about the past. The late afternoon sunshine glints on broad reaches of the Severn and on wide expanses of sand- and mud-flats, and the visitor is awed by the remote and melancholy grandeur of the cliffs of the Keuper Marl.

The rocks themselves are something of a geological conundrum. There is the same cuboidal weathering as at Wainlode Cliff and bands of Tea Green Marls occur at the top of the Red Keuper Marl —but there are also peculiar differences. There are distinct bands of green occurring at lower levels and even pieces of red rock can be picked up which have a core or centre of the green ferrous condition. Some geologists believe that this local reduction is due to bacterial activity. Often it radiates from a centre and often occurs in what may be an ancient crack. The organisms could have been awakened by access of moisture, after being encysted during arid conditions.

The geologist, of course, does not have a clear and easy rock 'blueprint' to read. Rather, his task is that of fitting together a gigantic jigsaw puzzle which is not only in three dimensions but whose 'pieces' have been scattered by movements of the earth's crust and warped and weathered by geological time measured in millions of years.

The rocks at Westbury-on-Severn are reassuringly legible in one respect, however. They are what is known as 'conformable', which

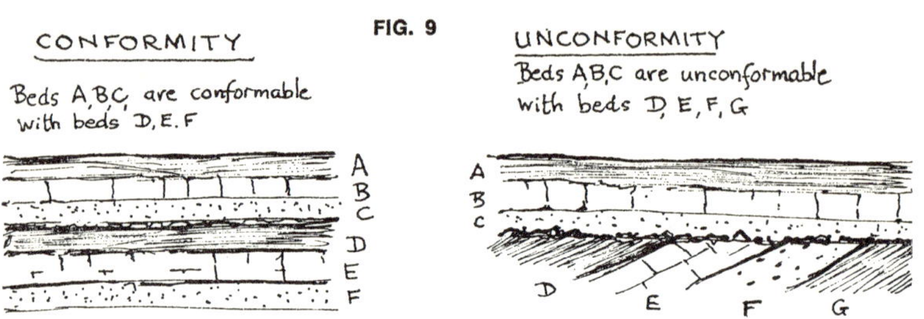

means that they lie in the same plane. The two diagrams at Fig. 9 show the geological distinction between conformity and unconformity.

THE RHAETIC BEDS AT WESTBURY

The Keuper Marl rocks at Westbury dip to the north-east so that the overlying Rhaetic rocks are brought down to the water's edge at the next point or 'headland'. Figure 10 shows how the beds of three distinct ages—Keuper, Rhaetic, Jurassic—all lie conformable.

It is best to look at Garden Cliff both from below and from afar; from below to see those extraordinary bands of Tea Green Marl (at least five of them in most places), from afar to see the alarming rapidity with which the top of the cliff is eroding. Large amounts of scree from this erosion have collected at the foot of the cliff and

FIG. 10

WESTBURY ON SEVERN   Triassic and Rhaetic rocks all conformable and dipping at 9°

the top of the cliff seems to be disappearing at the rate of several feet per year.

Notice, too, that the muds of the River Severn are red in colour, having been transported from the Triassic rocks further up the river. Other things to see as you go along the base of the cliff are the bands of thick sandstone which have fallen from the cliff top. This type of rock is called the Pullastra Sandstone of the Rhaetic series, being named after a fossil called *Pullastra* (Figure 12).

These flat sandstone slabs show ripple marks which were made on the sea shore in the Rhaetic lagoons about 190 million years ago. It is amazing to think that before one's eyes lie ripples and trails as they were left by the tides and by worms and other organisms on the shores of the Rhaetic sea. And yet, if we turn our glance to recent ripple marks on the tidal sand flats of the Severn, we must realise that these, too, would be preserved for millions of year if they happened to be covered quickly by rapid deposition, or a sudden sandstorm at low tide.

FIG. 11

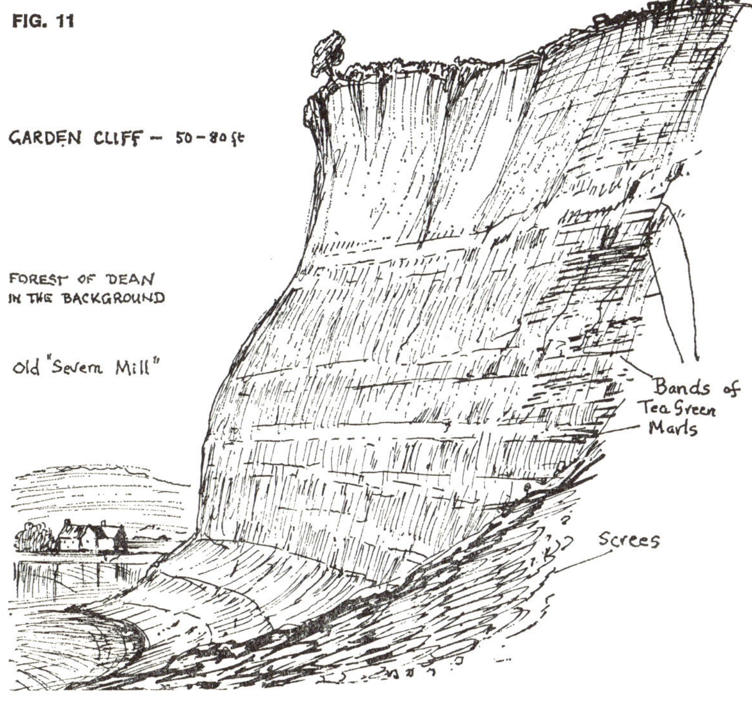

FIG. 12

The Pullastra Bed is a hard dark grey micaceous sandstone forming a conspicuous cliff feature. The fallen slabs form a natural protection to the foreshore.

The Rhaetic beds at Westbury are particularly worth close examination because it is generally accepted that they provide the best exposure of these particular rocks anywhere in the British Isles.

Figure 13 shows the general succession of the Rhaetic and part of the Lower Lias—the lowest division of the great Jurassic system— and is a very simplified version of the rocks the reader will see. The simplification is necessary because the Rhaetic beds are a series of thinnish strata laid down under peculiar conditions in Britain, although much thicker elsewhere. In Britain, they represent a lagoonal (or enclosed sea) transition stage between two great systems, the Trias (previous desert conditions) below, and the Jurassic (marine conditions) above.

As at Wainlode Cliff, the observer will notice that the cliff section shows a sudden change from the red rocks of the desert to the black shales of the Rhaetic, a stagnating lagoon phase as the Rhaetic sea slowly invaded the desert. The shales are black because ferrous sulphide has been produced by the action of bacteria on decomposing vegetation.

Marine reptiles and lung-fish congregated in these lagoons and a part of these Rhaetic bands, known as the 'bone bed', forms a rock cemetery which is packed with the fossilised fragments of fishes and reptiles. The bed is only a few inches thick, because deposition in these lagoons was so slow. Desert conditions had previously existed there, so there would have been no rushing rivers to bring down masses of debris, or to pile up sediment which would later form thick bands of rock. Only gentle currents winnowed the deposit, carrying lighter material away and concentrating the heavier debris, such as bones and phosphatic nodules.

This narrow bone bed is highly pyritised (the rock gleams with pyrite crystals) and packed with fish and reptile teeth and with nodules of black phosphate which are actually fossilised reptile

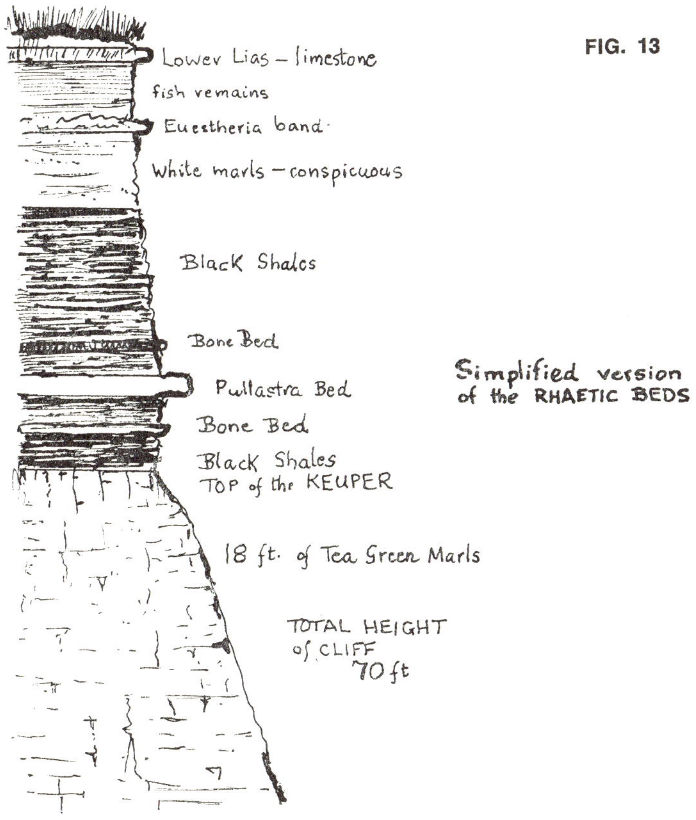

dung—amazingly well preserved and occurring in such profusion that they have even acquired a technical name, 'coprolites'. Other discoveries which could be made in the bone bed are Ichthyosaurus vertebra or perhaps the jaw of a lung-fish. This lung-fish, called *Ceratodus*, is particularly interesting because it belongs to that evolutionary stage when creatures living in water started to adjust themselves to living also on land, thereby paving the way to the development of creatures living entirely on land.

At that time a certain group of fishes used their swimming bladder also as an air bladder, that is, as a kind of lung to enable them to breathe in air while travelling from one pool to another. Thus, if pools dried up, they could survive, while those which lived exclusively in water would die out. (It must be noted here that amphibia had evolved from another group of fishes millions of years earlier, late Devonian period.)

FOSSILS OF THE 'BONE BED'

Figure 14 depicts a variety of bone-bed fossils, some of which can occasionally be found just a little above or a little below the actual bed. The sketch is worth careful study for you will only be lucky enough to make some good finds if you know what you are looking for. Notice particularly the shape of the Ichthyosaurus vertebra—rather like a cotton-reel.

Fossils are rarely found in the black shales and this dearth of relics of former life is evidence of the stagnating conditions in which the shales were deposited. Because of their resemblance to leaves of black paper, they are sometimes called 'paper shales' and some will be seen to have been streaked yellow by an efflorescence of sulphur minerals.

Reverting to the Garden Cliff, it will be noted that although the Pullastra Sandstone forms a conspicuous band of hard rock not far from the cliff top, the dip of nine degrees brings it right down to shore level where a grand platform overhangs the river. What has happened here is that the river has eroded the soft black shales underneath and left the sandstone jutting out as a platform.

Towards the top of the cliff can be seen a conspicuous band of white limestone. This marks the top of the Rhaetic series and is followed by thin bands of creamy-coloured limestone which form the very base of the Lower Lias which is, of course, the base of the Jurassic rocks.

FIG. 14

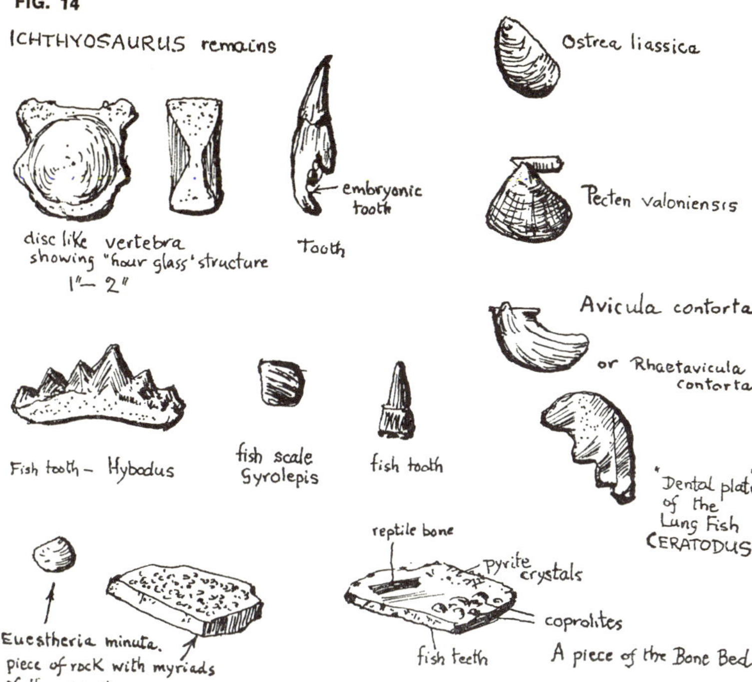

One other feature of this area, only occasionally visible at very low tide, is a peat bed—a very young deposit indeed compared with the Rhaetic rocks. Walking back along the cliffs to the old 'Severn Mill', you will see small cliffs of mud recently formed by the Severn, and a few feet below the grass is a layer of peat denoting the presence of a buried forest. The trees of this particular forest were birch and pine, and deer antlers have been found in the bed.

Obviously, when this forest was flourishing, the level of the River Severn must have been much lower than it is today. At that time, probably around 6,000 BC, England was still joined to the Continent and the Severn entered the sea somewhere between Bristol and Cardiff. Then, as the melting of the polar ice-caps in the last Ice Age was finally completed, the sea level rose, forming the English Channel, invading the rivers and flooding the nearby countrysides.

CHAPTER 4

The Severn Bridge

The Severn Suspension Bridge, situated about eight miles upstream from Avonmouth and spanning a mile of water between Aust Cliff and the Beachley Peninsula, is the seventh largest bridge in the world. It was opened by Her Majesty the Queen on 8 September 1966, and the cost of its construction was a round figure of £8 million.

Magnificently outlined against the sky, it is an artefact of sweeping grace, of light and seemingly effortless power, yet it supports a roadway from towers rising 450 feet above the river and carries a span 3,240 feet in length. It is a truly noble example of the type of architecture which can today express itself in high-grade steel, and a feat of engineering for which geologists will never cease to admire the men who dared to achieve it. For to them, the difficulties seemed well-nigh insuperable, so widely disparate in nature are the rocks upon which it stands and so numerous are the difficulties arising from their varying natural architecture. Topographically, too, the site presented a number of major problems, and when the engineers first began their daunting task in 1961 James Morris, recording his reactions in the *Guardian*, wrote:

> It is not an easy sort of place. The Severn here is more than a mile wide. The tides run faster than anywhere else in Britain and they rise and fall more than forty feet. The river is turbulent with shoals and eddies, convulsed at some seasons, viscous at others with millions of elvers wriggling their way to fresh water. When the water is low acres of glistening mud are revealed, morasses of harsh water grass, labyrinthine rivulets. Navigation is so tricky that in the old days many a foreign captain refused to load at Gloucester and the three little motor ferries that now run from Aust to Beachley struggle across the currents, when the tide is favourable, with infinite labour and circumspection.

For more than sixty years before the bridge was built, both sides of the Severn at this particular place had been the rendezvous of geological parties, who were drawn to it by the great variety of rocks offered by nature within so small an area. Why, then, did

THE SEVERN SUSPENSION BRIDGE FROM AUST CLIFF

the engineers choose so difficult a site? To understand this one must first appreciate the nature both of the traffic problem involved and of the rocks on the bed and sides of the river.

The traffic problem was to select a crossing-place on the Severn somewhere between Bristol and Gloucester to relieve the already appalling congestion at Gloucester and to cope with the still heavier traffic which completion of the new motorways would bring to the area.

Map 3 overleaf shows that important lines cross the river at two places, Sharpness and Beachley, while Map 4 shows in detail the nature of the rocks at Beachley. To appreciate the significance of these structure lines, the layman requires some explanation of the term and this can best be afforded by imagining Britain, not as an island, but as a region in the Northern Hemisphere some 600 million years ago. It was then something like a cracked pavement of hard, crystalline rocks and when sedimentary rocks were later laid down on this foundation the cracks and ridges still showed through or had some influence on the strata formation. This ancient structural line is known to geologists as the Malvernian or Malvernoid axis and is made visible on the surface by the alignment of the Malvern Hills and May Hill. The line crosses the Severn at Sharpness and continues on to Tortworth and Bath, where it has its influence on the famous hot springs.

C

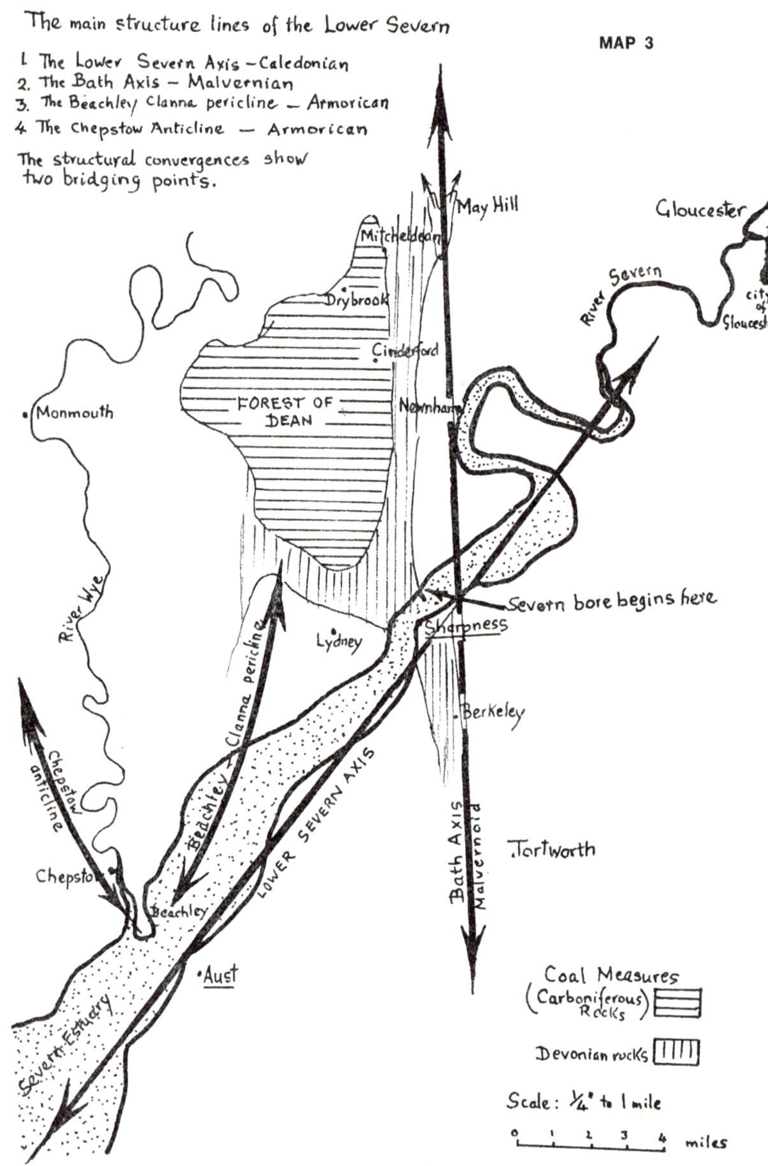

STRUCTURE LINES OF THE LOWER SEVERN

From time to time during the last 600 million years there have been periods of earth movements which have taken up alignments along certain directions. For example, it can be seen in Map 3 that the Lower Severn Axis is such a line and this direction, running NE to SW, is called 'Caledonian'. Other structure lines shown on this map are the Chepstow anticline—an upfold which causes a ridge—and the Beachley-Clanna pericline, which is a dome-shaped fold. The net result of structure lines is that, very often, older and harder rocks are thrown up near the surface, either as hard outcrops or as the result of the upheaval of rocks lying above them.

Looking again at Map 3, it can be seen that the Lower Severn Axis and the Bath Axis cross at Sharpness. This brings the Devonian rocks to the surface and, as they are harder than the rocks farther north along the river, they outcrop in red cliffs at Gatcombe and Sharpness. There, in 1874, a railway bridge was built, but a few years ago it was smashed by a barge which got out of control.

The Severn Bridge could have been built at Sharpness but possible traffic congestion and its greater distance from Bristol made it less convenient than Beachley. At Beachley the two structure lines shown on the map play some part in bringing the hard limestones of the Carboniferous rocks to outcrop not only on the river bed but in small cliffs to the west of Beachley. In fact, there are islands of these limestones almost in the middle of the river. What better geological choice, then, could be made!

There was also the added factor that by bridging the Severn at Beachley the appalling traffic congestion at Chepstow would be relieved. Furthermore, the Vale of Berkeley from Thornbury to Gloucester is thinly populated and likely to become a 'target area' for the overspill populations of Birmingham and London. All in all, Beachley stood out as the obvious place for the bridge.

THE ROCKS ON THE BEACHLEY SIDE

Further reference to Map 4 will show that Carboniferous limestones outcrop in a wide area to the south-west of the Forest of Dean and that the River Wye has cut through these rocks at Chepstow. The Chepstow anticline must have some influence on the outcrop of the limestones on the Beachley peninsula, and so must its convergence with the Beachley-Clanna pericline, for both meet on the ancient structure line of the Lower Severn Axis.

Along the foreshore west of the ferry landing-stage and under the bridge, the hard limestone can be seen outcropping. Although deeply fissured in places and well jointed, it was obvious that these rocks would form a good base for the bridge. In fact, the main problem in building the bridge was to understand the nature of these basement rocks.

Close inspection of the rocks towards Beachley Point reveals fissures filled with a hard breccia, i.e. angular lumps of the limestone cemented together naturally. This is the Dolomitic Conglomerate which betrays an ancient landscape surface.

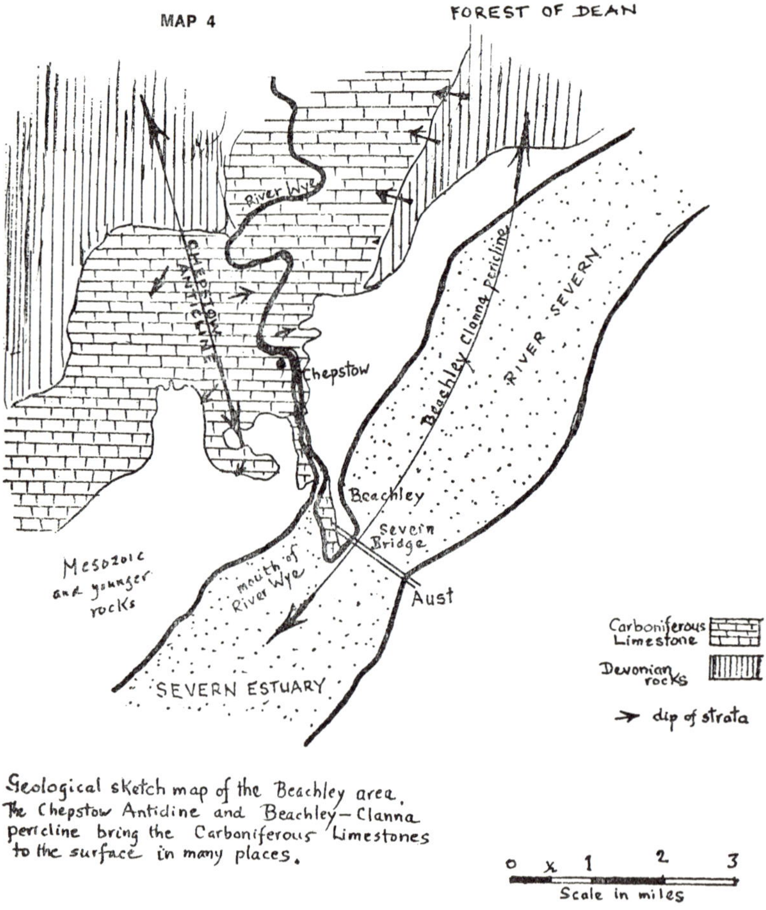

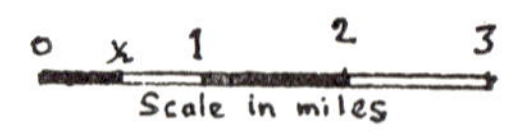

Geological sketch map of the Beachley area. The Chepstow Anticline and Beachley—Clanna pericline bring the Carboniferous Limestones to the surface in many places.

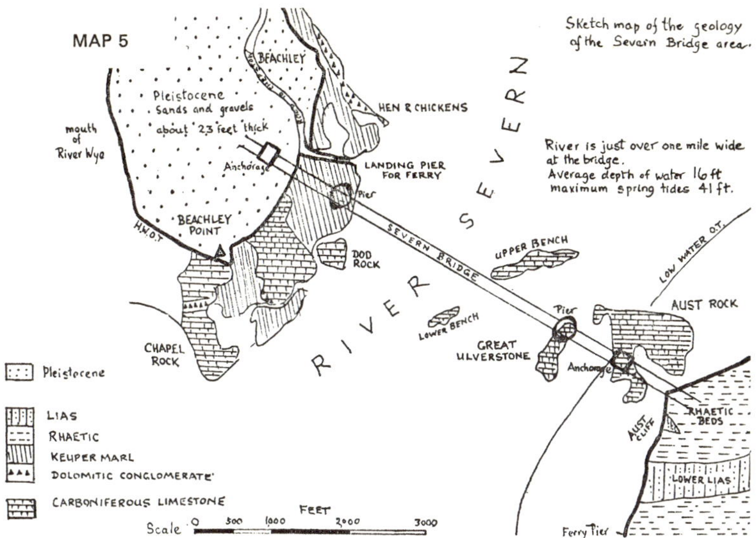

It can also be seen that the limestones are folded into ridges
(anticlines) and into small valley downfolds (synclines), both factors
which added to the complexity of the task of finding the basement
rock in which to sink the piers of the bridge.

The limestones are grey and are classified as 'dolomitic' (containing a proportion of magnesium carbonate as well as calcium
carbonate) and numerous fossils can be seen on the smooth surface,
particularly the columnals of ancient sea lilies called crinoids. In
places, corals can be seen and these are useful in identifying the
limestones as being of Carboniferous age.

The geological sketch (Map 5) shows that the Beachley Point
peninsula is covered with sands and gravels to a depth of some
twenty-three feet. These are of recent geological age (Pleistocene)
and were probably related to glacial and interglacial periods when
the Severn flowed in other areas and at higher levels.

It can, therefore, be seen that the Beachley anchorage concrete
base had to be dug down through these soft sands and gravels in
order to reach the basement rocks. In fact, the steel tubes here go
down some sixty feet, and getting them into position presented no
problems for the engineers as the site was on dry land.

Map 5 also shows that the areas of limestones have intervening
regions of the Keuper Marl, the same rock as seen at the base of

the Aust Cliffs on the opposite shore. These red rocks form cliffs in a few places near Beachley Point.

After the Carboniferous limestones were laid down some 300 million years ago they became folded, eroded and formed an ancient land surface in the period of the Triassic deserts 200 million years ago. The ancient valleys trending in a NW direction were the first to be filled up with Triassic sediments—the Keuper Marl. This presented a problem to the engineers who bored down for the site of the Beachley pier to support the great steel tower. After going down thirty-four feet, some hard mudstones were encountered, but not the good hard limestones. Nevertheless, they proved to be Carboniferous rocks by the nature of the fossils found in them and the engineers considered that these steeply-dipping beds presented a strong enough structure for the Beachley steel tower for the suspension of the bridge. No doubt these mudstones will prove to be sufficiently strong but it is the writer's opinion that the boring should have gone even deeper in order to get down to the greater security of limestone rocks. Even the monks in the Middle Ages recognised the greater security presented by using limestone as a foundation, and there is a ruined medieval chapel, St Twrog, on the limestone rock known as Chapel Rock. Incidentally, it is rather risky trying to reach this particular rock, as there is a deep channel in one of the fissures which soon becomes impassable with the incoming tide.

Even in midwinter, quite a pleasant afternoon can be spent here in one of the miniature coves collecting calcite crystals from the limestones, examining the small anticlines and various structures exposed on the foreshore. It is also quite a good place to demonstrate how limestones *can* be folded, for it is most difficult to imagine this when merely confronted with a hard lump of rock in the hand.

The sketch in Figure 15 shows the Carboniferous limestones plunging away under the Severn.

THE AUST CLIFF SIDE OF THE SEVERN

These cliffs are visited by geologists from all over Britain because the best display of Rhaetic rocks in the whole country lies towards the top of the cliffs and contains the highly fossiliferous reptile 'bone bed'.

The cliffs reach a height of over 140 feet and are mainly com-

FIG. 15   Small cliffs of Carboniferous Limestone, about 20 ft high, at Beachley Point. – well jointed and fissured. A small upfold or anticline can be seen in the foreground In some of the fissures are masses of the Dolomitic Conglomerate which forms the base of the TRIAS.

posed of Red Keuper Marl, but about eighty feet up the colour changes to the greenish tinge of the Tea Green Marls (not to be confused with blotches of green marl within the Red Keuper beds!).

This Keuper Marl is really a compact of very fine silica dust laid down by winds blowing across the Triassic deserts some 200 million years ago. The maximum thickness of the red rocks here is about 200 feet but they reach a total thickness of 2,000 feet in Cheshire, where they contain valuable salt deposits, indicative of vast salt lakes in ancient deserts.

Visitors to the site of the Severn Bridge may find themselves becoming quite alarmed at the behaviour of these red rocks, as on sunny afternoons the noise of lumps falling down can be heard continuously, and the lumps themselves can easily be crumbled into dust. Hardly the kind of rock on which to build a bridge, one might say, so what is it that is holding the rock together? One theory is that it is a heavily compacted rock of extremely fine particles merely held together by physical bonding. Once it is 'inside' the cliff it remains a hard rock but immediately it is cut into and exposed to the weather it quickly disintegrates by this curious process of cuboidal crumbling.

The cause of the disintegration stems from the fact that clay minerals have a great attraction for water, and when the surface of the rock is exposed to water it swells and when it subsequently dries out it shatters at or near the surface. This is a problem which had to be considered when building the approach roads on the Aust Cliff.

High up on the cliff can be seen a sudden change from the Tea Green Marls to deep black shales which are about eight feet thick. These are known as the Rhaetic beds, which were laid down in shallow lagoons when seas began to invade the Triassic deserts about 190 million years ago.

Above the black shales are thin bands of grey argillaceous limestones with a band of yellow clay intercalated. These are the Upper Rhaetic beds and indicate the onset of marine conditions. Finally, at the very top of the cliffs, but not continuous along the cliff edge, are shales with thinly-bedded limestones of the Lower Lias.

Between the old ferry pier and the new bridge, small headlands jut out and it is here that Rhaetic and Lower Lias rocks can be seen brought down to a lower level. Although this could be due to faulting, it is also possible that they are landslipped blocks which have merely slipped down because the top part of the cliff collapses when the softer rocks at the cliff base are eroded away. It can be seen that the Lower Lias limestones at the cliff top are brought down to a lower level and being more resistant to erosion are thereby responsible for the promontories.

The faulting has also changed the inclination of the beds to a different angle and direction (northerly) and the consequence is the appearance of small springs in the fault zone. The surface water percolates through the limestones at the top and is released at the junction of a thin bed of clay with the fault. This causes a profuse mass of vegetation and, as the water coming from the limestone rocks at the top of the strata is rich in calcium carbonate, this vegetation—mostly mosses and grass plants—is petrified into weird stone 'plants'.

On the foreshore are extensive mudflats, which are colonised by the salt-loving *Spartina townsendi* grass. About fifty years ago some seeds of a South American grass plant came off a ship at Southamp-

FIG. 16

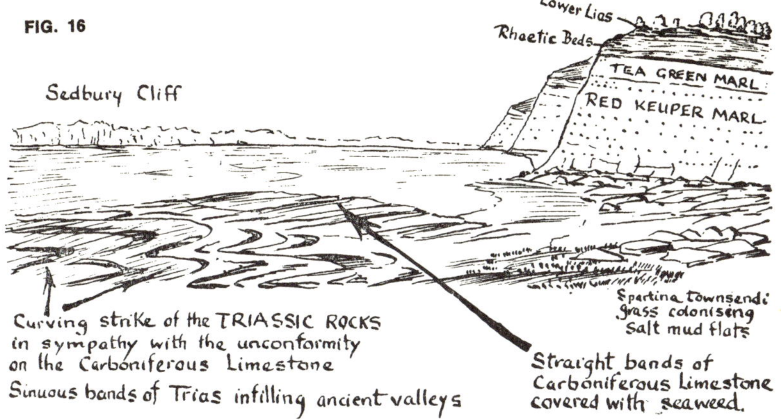

ton Docks and crossed with an English species. Later came a related new species which spread all round the coastal mudflats of Britain.

Out across the waters of the Severn are seaweed-covered rocks, some of which show remarkable curving lines. Figure 16 shows that they represent the very base of the Triassic rocks following the ancient valley system developed on the surface of the Carboniferous limestone rocks.

AUST ROCK

Aust Rock (see Map 5) is a planed-off surface of the limestone upon which rests the Aust anchorage for the bridge. The engineers found the rock was rather fissured and had to go down ten feet to get a secure hold. The Aust pier, which carries the weight of the 400 feet-high tower, is on Ulverstone Rock, and only a four-foot pit was necessary here as the limestone on this rock was very hard.

Near the bridge it can be seen that the whole structure of the cliff is that of an upfold or anticline, the crest of which is just by the bridge at Aust Cliff, as shown in Figure 17.

FIG. 17

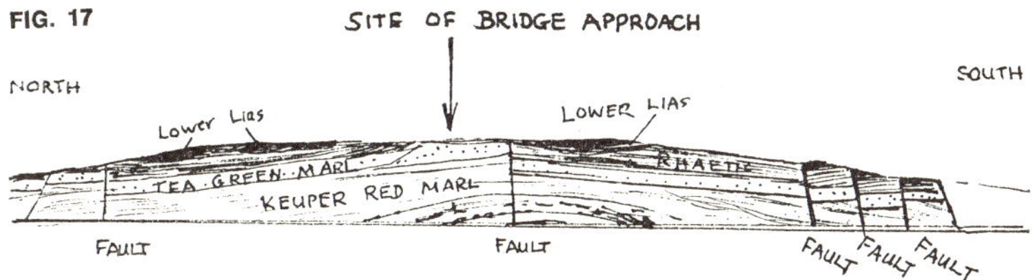

The strata gradually become horizontal towards the cliff top, and Professor Whittard, who made a detailed survey of this area in 1949, came to the conclusion that it was a 'compaction structure'. This implies that the first series of Triassic beds (the Keuper Marl) were laid down on an uneven surface and these structures were developed when the rocks 'settled down' or became compacted.

Right on the crest of this anticlinal structure and close to the bridge it can be seen that there are streaks of white to pink rock. These vertical bands are alabaster, a variety of the calcium sulphate mineral gypsum, and branching off horizontally are thin white streaks of another form of gypsum called Satin spar. This white mineral shows a beautiful silky lustre.

The alabaster fills in fissures in the Red Keuper Marl, but in many places the sides are curved and show a pattern comparable to the contraction cracks which can be seen in the dried-up muds of the foreshore. It is believed that the gypsum is derived from saline lakes in the ancient Triassic deserts, a feature often seen in salt lakes today in the Middle East. This material could be used for making plaster of paris, but the deposits are not thick enough for commercial exploitation.

THE BRIDGE APPROACH ON THE AUST SIDE

The motorways approaching the bridge on the Aust side are excavated through the Upper Lias limestones, the Rhaetic beds and down through part of the Keuper Marl. Although Figure 18 does not show the real nature of the problem it does show how the cliffs of Keuper Marl provide stability for the approach road.

Much research has gone into the problem of rock mechanics in the red Triassic rocks because so many roads of the Midlands are cut through these rocks and the problems which arise as the result of their weathering demand careful study.

It is perfectly sound 'rock mechanics' to build the approach road on the Keuper Marl, but this rock must not be disturbed too much. Once it is disturbed, weathering takes place and all the disintegration factors of dehydration and shrinkage begin to operate. The correct angle of slope of the great embankments on the approach roads had also to be carefully planned and all slopes were grassed over very quickly to prevent erosion.

Figure 19 on page 44 will give the reader an idea of the rock structures encountered in building the Severn Bridge. It should be remembered that it is very difficult to find out all details of the structures in the limestones unless a large number of borings are made through the seaweed-covered rocks.

Finally, visiting amateur geologists studying the rocks at Aust Cliff should always bear in mind the correct sequence of the strata in this area. Most of the specimens they will find will be 'out of order' chronologically because they will probably be taken from

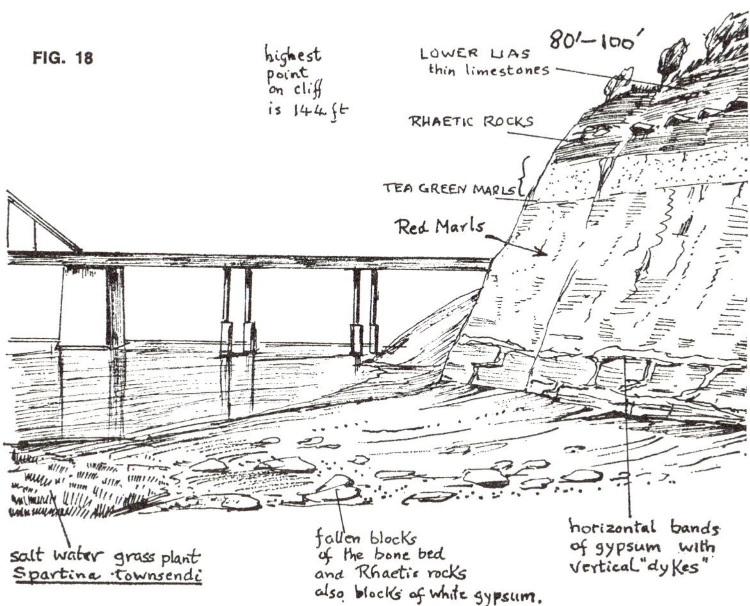

APPROACH TO BRIDGE ON THE AUST CLIFFS

the numerous blocks which have fallen to the foot of the cliff.

The rocks which they are most likely to pick up are as follows:

(i)   Pieces of the 'bone bed'.

(ii) Pieces of the Lower Pecten Bed (black fossiliferous limestone).

(iii) Slabs of arenaceous limestone.

(iv) Pieces of hard cream-coloured limestone.

(v) Larger slabs of the Lower Lias limestones with the bedding planes covered with a kind of oyster fossil called *Ostrea liassica*.

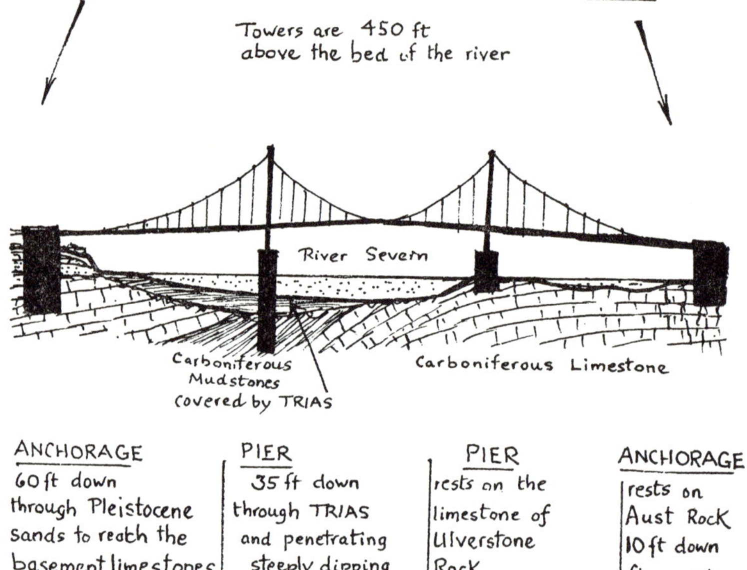

FIG. 19   THE ENGINEERING GEOLOGY OF THE SEVERN BRIDGE

THE AUST CLIFF SUCCESSION

The following is a list of the strata in the order in which they can be observed from the top of Aust Cliff down to the base.

LOWER LIAS	=shales and thin bedded limestones
	=Cotham Marble, 6 in thick
	=yellow clay with limestone bands, about 4 ft thick
UPPER RHAETIC	=grey argillaceous limestone, 2½ ft thick
	=yellow thinly-bedded argillaceous limestone, 4 ft thick
	=greenish black shales, 1 ft thick
	=hard grey limestone called the Upper Pecten bed, 1 ft thick

<table>
<tr><td>LOWER RHAETIC</td><td>=black shales, 8 ft thick
=hard pyritised limestone called the Lower Pecten bed, a few inches thick
=hard fissile paper shale, about 8 in thick
=the famous 'bone bed', 1 to 4 in thick
=Grey or Tea Green Sandy Marl, 3 ft thick
=hard sandy bed, 1 ft thick</td></tr>
<tr><td>TRIASSIC ROCKS</td><td>=Grey or Tea Green Sandy Marl, 18 ft thick
=Red Marl, 52 ft thick
=gypsum series, 25 ft thick
=Red Sandy Marl, 20 ft thick</td></tr>
</table>

The red beds of the Triassic rocks have not revealed any fossils, but the writer has found leaf imprints in the Keuper Marl of Wainlode Cliff, near Tewkesbury. Of the fossil invertebrates found in the Rhaetic and Lower Lias beds, bivalves are the most common. Next come arthropods and a few plant remains. For the identification of fossils, the best book is *Mesozoic Fossils*, published by the British Museum of Natural History, price 12s 6d.

CHAPTER 5

The Severn Bore and Hock Cliff

The one thing that almost everyone knows about the River Severn—and often it is all they do know about it—is that it has a bore. Just what a bore is and why the Severn has one when other rivers, including even the mighty Thames, do not, is less well known, though it can be simply explained. The behaviour of a river is determined by the nature of the rocks it flows over and between, and the Thames flows over very different rocks from those of the Severn. In its tidal part, the Thames flows over a clay plain with no hard bands of rock to complicate its manner of flowing. The River Severn, too, flows over and alongside a type of clay but in that clay and above and below it are many other bands of different and sometimes harder rocks. Its passage is further complicated, and navigation upon it made the more hazardous, by hard bands of limestone which outcrop through the clay, as well as by shifting shoals of sands which occur in many of the estuarine stretches of the river. In fact, the Severn probably has more sands and muds than the Thames because the Severn Basin is in softer sedimentary rocks and also receives a heavy run-off from the high rainfall of the Welsh mountains.

Let us now take a closer look at the rocks which hold the 'secret' of the Severn's bore. Hock Cliff, some five or six miles up-river from Sharpness, is the best place to observe them. Here, near Fretherne, the cliffs are of Lower Lias clay, and remember that, to a geologist, clays, sands and gravels are 'rocks', just as much as more obvious rocks such as limestone and granite. Remember, too, that the Lower Lias is the oldest band of the rocks of the Jurassic system and was laid down after the rocks of the Triassic system—so that the rocks at Hock Cliff, representing a period approximately 170,000,000 years ago, are younger than those of Aust.

Hock Cliff is of outstanding geological interest because it is rare to find a cliff or quarry-face of clay with the bands of strata so clearly visible. Usually, clay either crumbles away or is quickly grassed over. We have already learned that clays denote a muddy sea, rather deep, but in the case of Hock Cliff it is reasonable to suppose that the sea there was not as deep as in the area of North Gloucestershire and near Evesham. There the Lower Lias clays are 960 feet thick, whereas at Hock Cliff they are only between 200 to 300 feet thick. One should not infer from this, however, that a thick

FIG. 20

HOCK CLIFF

FIG. 21

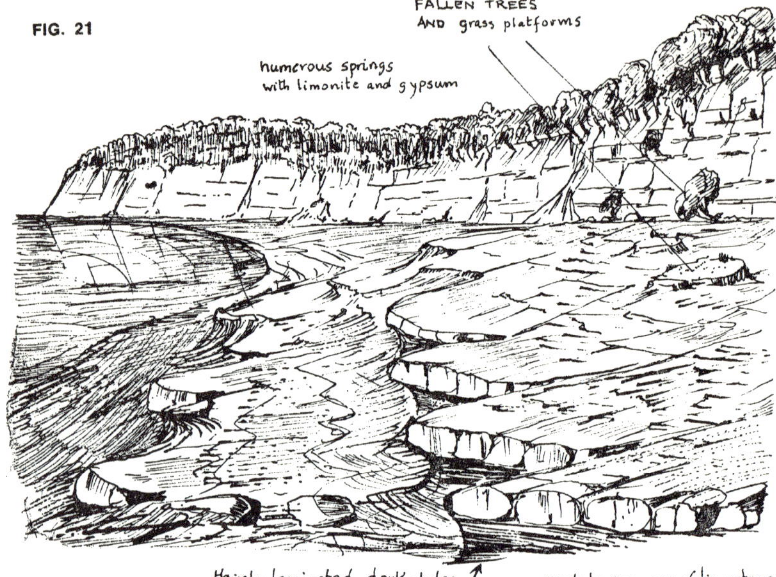

MASTER JOINTS IN THE LOWER LIAS AT HOCK CLIFF, FRETHERNE

deposit invariably indicates a deep sea because this situation can also occur with a constantly sinking sea bed.

Hock Cliff is no simple example of a clay cliff and a river flowing placidly over a clay bed. The Mendip Hills, the eroded remnants of a much larger range of limestone mountains, are not far away and rivers running off that range of mountains deposited the erosion debris in the muddy seas below. Hence, in the clay of Hock Cliff there are recurring bands of limestone about a foot thick. The number of bands to be seen varies with the height of the tide, but an observer in 1901 reported seeing as many as twelve. The writer has seen three bands in one part of the actual cliff structure, increasing to five further up the river, and has noticed that the recurring pattern of the bands is repeated on the foreshore, which goes down almost in limestone 'steps' set in the clay.

The number of 'steps' visible again varies with the height of the tide and the condition of the water, but one constant and almost uncanny feature is the way in which the limestone bands all 'strike' in the same direction. This consistent similarity in the direction of the rocks is repeated in the cliffs above where the bands of limestone nearly all project from the softer clay in whole lines of triangular shapes at an angle of about forty-five degrees. Taken together, this remarkable consistency, evident in both cliff and foreshore, makes it reasonable to assume that the pattern is repeated on the *bed* of the river, and that there are bands of hard rock outcropping in the clay.

THE SOURCE OF THE BORE

And, looking at the water at low tide, further confirmation is to be found in the occasional darker folds in the water, as if it were running over ledges at those places. Furthermore, in those same places a higher level of 'rushing' noise can be distinctly heard against the background of the water's gentle lapping against the mud and stones of the foreshore. It is a microcosm of a sound, the thunder of a mighty waterfall represented in miniature. But it is also the 'secret' of our famous Severn bore, for here we are only a mile or two away from that part of Frampton Sand where the bore begins.

For a detailed, mathematical exposition of the causes of the bore, the reader is recommended to Dr R. A. R. Tricker's excellent book *Bores, Breakers, Waves and Wakes* (1964). Here only a simple geological exposition will be attempted, though mention must be made that part of Dr Tricker's explanation is that, where the bore begins, the Severn encounters a sudden rise or 'step' in its bed. And at Sharpness, only two miles further downstream, the width of the

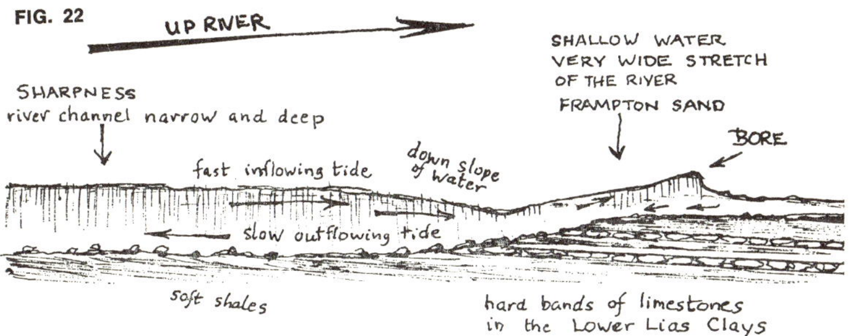

CONDITIONS FOR CAUSING THE SEVERN BORE
A FAST TIDE ENTERING A RIVER WITH A SLOPING BED.

D

river narrows rapidly to only one mile as against its two miles width at Frampton Sand. This constriction of the river is the second main reason for the bore and also has a geological explanation.

Figure 23 explains the geological significance of the area round Sharpness and shows a convergence of the main structural lines of the rocks which control the build of this part of Britain. These rocks are hard, and a river flowing through hard rocks remains narrow; only when flowing through soft rocks does it erode its banks and become wide. Thus the Severn is narrow at Sharpness whereas, alongside Frampton Sand where the rocks are soft, the river widens out, although there are here some dangerous bands of hard limestone rock which outcrop through the clay in 'steps'. It should be mentioned that Silurian rocks outcrop along the Severn at Tites Point near Sharpness. They consist of calcareous shales and sandstones.

When the incoming water piles up and is concentrated in the

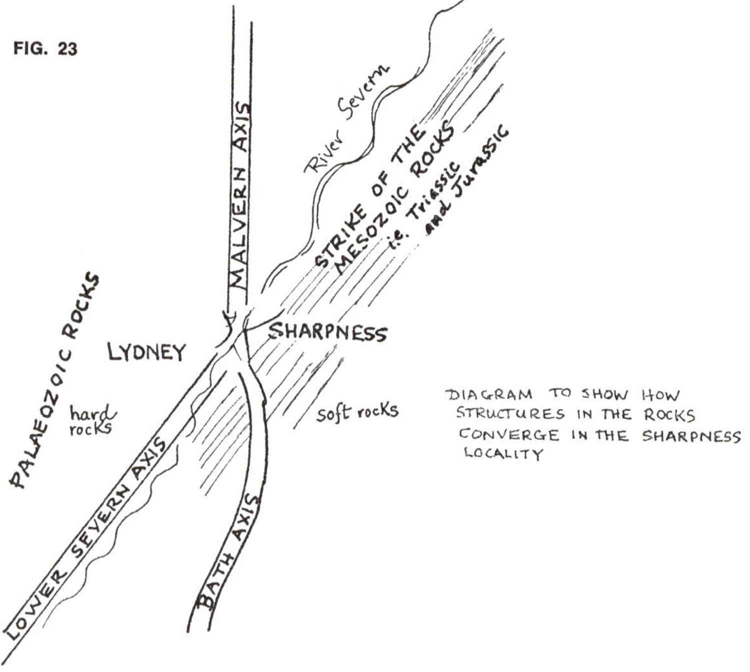

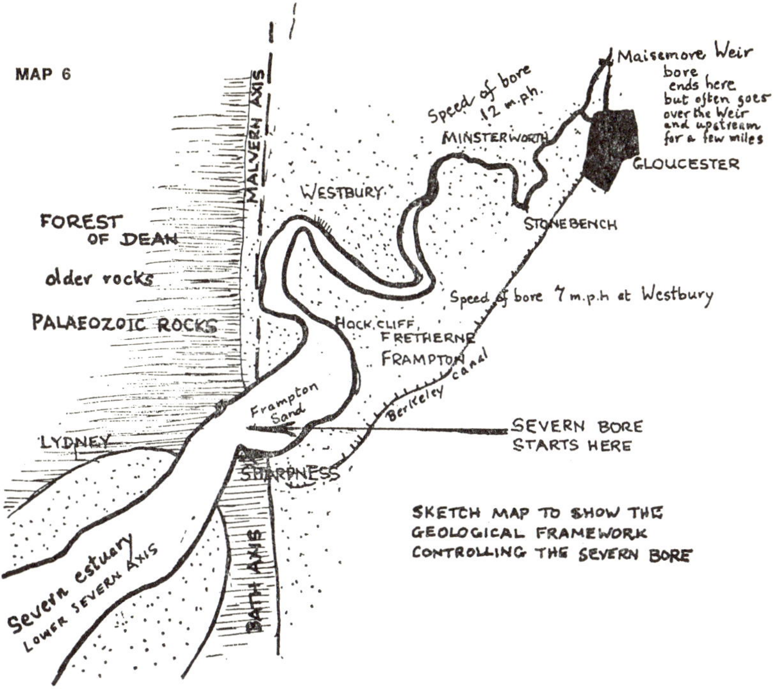

narrow part of the river at Sharpness the rate of flow is accelerated and the fast-moving water rushes on to meet the obstruction of a step of hard rock followed by the wide, shallow stretch of sand at Frampton. Thus is formed the bore beginning at Frampton Sand.

HEIGHT AND SPEED OF THE BORE

The height of the bore depends on the configuration of Frampton Sand which, in turn, is controlled by the master joints in the Lower Lias bedded limestones. The speed of the bore depends upon its height and the depth of the river water. In fact, there will be no bore at all if high-water at Sharpness is less than twenty-six feet, a situation which tends to occur at the Spring equinox, within about two or three days of a new or full moon.

Dr Tricker points out in his book that another factor affecting the speed of the bore is acceleration due to gravity and gives some fascinating mathematical formulae illustrating the state and speed of the bore in varying circumstances. The formulae may be somewhat daunting to the average reader who may prefer to settle for the two following useful pieces of information:

(i) A good definition of a tidal bore is that it consists of a body of water advancing up the river with the incoming tide and having a well-defined front which separates it from the slowly ebbing water into which it is advancing. That front takes the form of a wave or a series of waves.

(ii) From its inception above Sharpness, the bore increases in height and attains its maximum between Framilode and Stonebench. Thereafter, the height slowly diminishes towards Gloucester, although the bore travels beyond that city. It can be well observed at Minsterworth on the right bank of the river, or at Framilode and Stonebench on the left.

If you observe the bore from its point of origin, stay awhile to look at the rocks which cause it and to examine the types of fossils which help to identify them.

Those limestone bands are called argillaceous limestones because they are so clayey, the word 'argillaceous' merely meaning 'clayey'. Local farmers, on the other hand, simply, and just as accurately, call these stones 'claystones'.

For centuries men have taken advantage of having in these stones readymade ingredients for cement—finely ground clay and limestone—and, in some places, these limestone beds are known as 'cement beds', and have been quarried in the past to make hydraulic cement.

THE FOSSILS AT HOCK CLIFF

Hock Cliff is also a happy hunting-ground for fossils but it is a peculiarity of the area that, although fossil fragments can often be found clustered thickly together, there are many large expanses of rock without trace of a single fossil. These are where the presence of highly destructive iron sulphide in the clay has tended to discourage the survival of fossil remains.

Where fossils have survived the following are the most plentiful and the most interesting:

(i) *Gryphaea arcuta*, a curiously-curved oyster whose fossil shell, powdered and mixed with whey, was used in the Middle Ages as a cattle medicine. Fretherne foreshore, near Hock Cliff, is internationally famous as a good place to find this fossil.

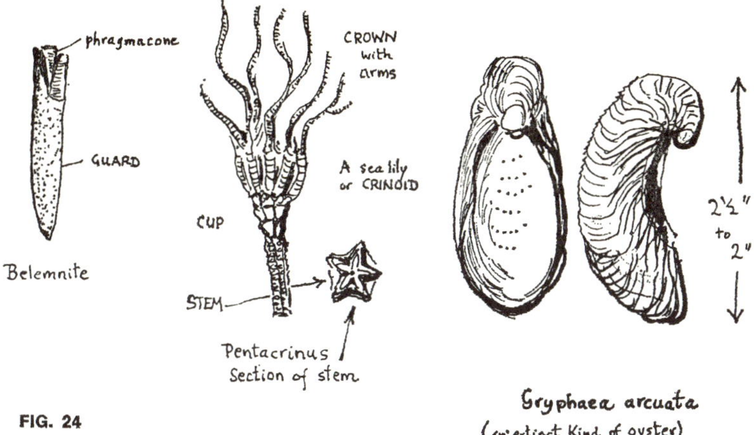

FIG. 24

(ii) Belemnites: these look like slender bullets but are really the hard calcareous 'guards' of an extinct creature which looked like a squid.

(iii) The limestones abound with the columnal fragments of the sea lily, *Pentacrinus*: these exquisitely-beautiful five-pointed star shapes are a cross-section of the main stem and the dozens of dots which may sometimes be found near the stars are the cross-sections of the dozens of smaller branches or rootlets from the main stem.

(iv) Ammonites: these can be found in the shales and are often pyritised, i.e. the fossil has been replaced by pyrites (iron sulphide).

CHAPTER 6

The Severn Terraces

The Severn shares a common feature of most rivers in that it has natural terraces. Beyond those treacherous, gleaming mud flats and the rocks which jut out of the water like the teeth of legendary reptiles, there are, further up the river, pleasant fields and meadows which were all part of a wide erosional valley carved out long ago by the river into a remarkable series of green 'terraces'. For in those far distant times the Severn, like most other British rivers, was much wider, higher and mightier than it is now—the result of the melt water of the ice-sheets and the heavier rainfall during inter-glacial periods.

This 'terrace' region is a quiet backwater, a softly pleasant land rather like the Constable country of Essex and Suffolk, where fertile fields are interspersed with sleepy villages—Apperley, Tirley, Norton, Deerhurst, Hasfield, Ashleworth, Sandhurst and Maisemore.

THE TERRACE VILLAGES

If you branch off from the Haw Bridge road you will get to Apperley, where the lanes are lined with typical houses of the plains, either in red brick or half-timbering. There is even an excellent example of an old cruck cottage at Apperley, conscientiously restored by Cheltenham Rural District Council, and a magnet for history students every year.

Remember that the kind of housing which occurs most frequently in any area is often a good pointer to local rocks since these usually provided the building materials for the poorer houses. This area is no exception and only churches and the houses of the wealthy have imported Cotswold stone in their fabric.

Look at the map and you will see that Apperley is very close to the Severn. Yet go through Apperley and you cannot at first see the Severn! But if there has been heavy rain up-country and the river is in flood you can look through the trees there and, surprisingly,

Village of Apperley — a cruck cottage
Most of the houses on the clay plains
are half timbered or of red brick.

find yourself gazing down on flooded meadows resembling an inland sea. The village itself, standing over 100 feet above the flood plain, is safe enough, but an abrupt descent will bring you to the White Lion Inn right by the river, and here the unpredictable Severn has been known to lap up to the level of the bars!

The White Lion Inn, APPERLEY. The village of Apperley is up on the hill terrace
100 feet above the river.
In the old days barges brought coal for the village, hence the name "coal house quay"

The White Lion served as a coal-wharf in the old days when barges came up the river and dumped the coal to be taken up to the natural terrace on which Apperley stands. Other pubs important in local history are at Haw Bridge and Ashleworth, also formerly useful as ferry crossings or coal wharves.

If you turn away from the flood plain and cross Haw Bridge you will come to the next village, Hasfield, and again it is obvious that this is also on a terrace, although a lower one than that at Apperley. Like Apperley, Hasfield is safe from the flood menace of the Severn, and this menace is a very real one, for the Severn floods much more rapidly than the Thames. When there are heavy winter rains in Southern England, much of the water merely sinks into the chalk rock surrounding the Thames Basin, whereas the hard rocks of Wales allow a rapid run-off, quickly bringing the River Severn into full spate. Hence those village settlements clustering on the terraces.

THE WOOLRIDGE TERRACE

Behind the village of Hasfield there is another high hill and, approaching the tiny settlement of Woolridge, it is easy to see that here is yet another terrace, a kind of remnant plateau. This particular 'flat' is also partly structural—there is evidence that it is an 'outlier' of Rhaetic rocks from which the surrounding rocks have been eroded away—but it is nevertheless still a river terrace carved out by the ancient Severn.

Look at the fields around and you will see that they are positively

A RIVER TERRACE   The village of Hasfield lies on a gravel terrace some

FIG. 27   80 ft high or 50 feet above the Severn flood plain.

gleaming with ploughed-up pebbles, some of them quite large. These smooth, round pebbles are 'rocks' quite alien to this district for they are of Bunter Sandstone, which is found in the Midlands and have been swept here by an ancient river mighty enough to carry along pebbles larger than a man's fist.

Professor L. J. Wills named this terrace the Woolridge Terrace (after the nearby village, *not* the famous geomorphologist, the late Prof. Wooldridge), and if you look across the Severn to Sandhurst you will see a similar terrace known as Norton Hill. But to geomorphologists it is not so much Norton Hill but more the Woolridge terrace once again because it is the same height, 250 feet, and was obviously formed at the same time and subjected to the same conditions.

FIG. 28

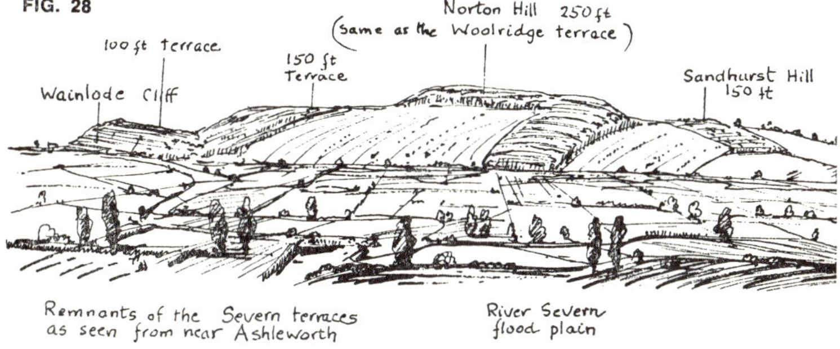

Geomorphology, it should be explained, is the study of the form of the ground, of the shape which a landscape has taken owing to relatively recent erosion and relatively recent and therefore 'superficial' deposits. A rapidly developing natural science, geomorphology already promises to achieve some independence both of geology and of physical geography but, while the boundaries still overlap, areas such as the Severn Terraces remain a source of endless fascination to this new kind of scientist, so abundant are the clues they offer to the varying behaviour of land and water.

To the tourist, the terraces offer panoramic views of breathtaking beauty. If you stand just above the village of Sandhurst and look out across the Severn you can see east to the Cotswolds, south to Gloucester Cathedral and far away north-west to the Malverns. And below you, the Severn glints among the trees and fields, appearing, disappearing, reappearing, constantly changing its character with the changing light of day.

And when you are up there, or on the Woolridge Terrace on the opposite side of the river, look out also for the erosion scars on the hillside, those ancient meander scars the river made when it was flowing at this much greater height, fed by the melting ice of the great continental polar ice-caps.

That was, very roughly, about 10,000 years ago, and within the last million years or so there have been several warm inter-glacial periods during which water from melting ice-sheets flowed south from the Midlands creating great rivers which spewed gravels from the Midlands all over the plains where we now see the Severn

terraces. These gravel terraces are thus related to periods when the ancient Severn was much bigger and had far greater erosive power than today and represent much longer intervals of time than the steps in between which relate to a period when the river was cutting down more rapidly.

And, incidentally, for those interested in archaeology, it is worthwhile keeping a lookout for the flint implements of Palaeolithic Man and the mammoth remains that are sometimes to be found in the terrace gravels.

As the gravels on the 250 ft-high Woolridge Terrace are obviously the oldest, the sequence of what has occurred in geomorphological time is, in this instance, the very reverse of what happens in geological time. William Smith, it will be remembered, proved that

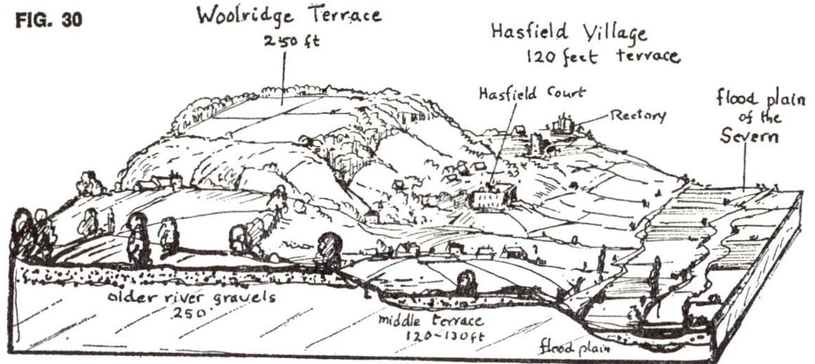

The Severn terraces between Ashleworth and Haw Bridge

when rocks are horizontal the older beds are those which are below and those on top are younger. Yet on the Severn terraces we have the older beds (of gravel) up above. But this is characteristic of the whole complicated geological history of the Severn, though neither geologists nor local farmers have any reason to complain about it. For geologists, the region is one of altogether exceptional curiosity and interest. As for the farmers, thanks to the conditions created by the Severn, they are farming good alluvial soil, and even the river's unpredictable flooding is no hardship to them because the network of terraces provides superb high-and-dry sites for farms and villages. The land is fertile and also easy to work because it is kept well-drained by the underlying sand and gravel. Finally, roads can run across the higher flats of land, so completely avoiding the flood plain.

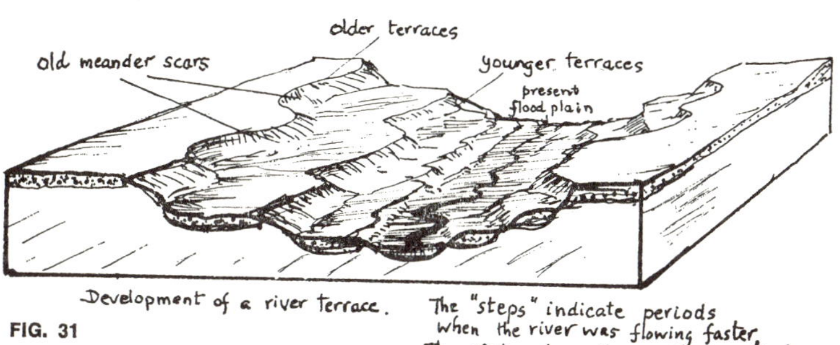

Development of a river terrace. The "steps" indicate periods when the river was flowing faster. The flats where it was flowing slowly

FIG. 31

Chapter 7

The Cheltenham Sands

'Cheltenham Sands' are not found only at Cheltenham, and the term applies equally to many other deposits of similar sands found scattered haphazardly over the Severn Vale. They are classified by geologists as superficial deposits and it is interesting to map both the areas where they are found and the locations where village settlements have existed for hundreds of years. It will be found that they dovetail very neatly if you go back to Saxon times.

When the Romans left Britain in the fourth century AD, the Saxon invaders began to settle in plain areas, which had been deliberately avoided by the earlier Neolithic and Iron Age tribes because they had no heavy tools capable of working the heavy clay soil. Instead, they kept to areas like the Cotswolds and the Chilterns which had much lighter soils and where the forest cover was thinner.

But the Saxons brought with them better ploughing implements and so were able to work the heavy Lias clays of the plains. At the same time they looked around for dry sites for their homes and these they found wherever there were deposits of sand.

This practice of seeking sand on which to build was pursued right up to the last century, and Cheltenham itself was, originally, merely a single street on a convenient patch of sand. Today, however, the city has grown so big that it has sprawled far beyond that original patch of sand and many of its houses are built on clay.

Anyone lucky enough to have a house on the sands in Cheltenham will be dry and his garden will be easy to cultivate, whereas if his neighbour's house on the opposite side of the road happens to have been built on clay his walls may tend to crack, window frames go slightly askew, chimney stacks get out of alignment—and he will probably always be complaining that his house is damp.

Obviously, then, anyone buying a house would be well-advised to find out the whereabouts and depth of these superficial deposits of sands in his neighbourhood. They vary widely in thickness from nearly zero in some parts of the country to a maximum depth of fifty feet at Charlton Kings, which may help to explain the great prosperity of this particular area!

The villages of Swindon, Gotherington, Churchdown (not Churchdown Hill), Bishops Cleeve, Alderton and Twyning all originated on patches of sand. If you visit them, look out for diligent gardeners in the older parts of the villages and notice the light sandy soils being turned up by the spade. The new housing-estates growing up round villages have, however, to put up with heavy clay which, no doubt, explains the occasional untended garden one sees on these estates.

FIG. 32

The shape of the village of the village of Gotherington is almost identical with the outcrop of the Cheltenham Sands.

GOTHERINGTON

[Lias Clay — 170 million years old
Cheltenham Sand ¼ million years old]

At Gotherington, by way of contrast, the present shape of the village reflects almost exactly the outcrop of the Cheltenham Sands, despite the fact that Gotherington has seen a dramatic rate of growth during the last few years. The explanation is that much of that growth has occurred by the process of 'infilling' between existing houses and spreading out further along existing belts of sand.

The sands are not only valuable as sites but also provide basic building material for the rapidly-developing areas around Cheltenham. And when sixty tons of sand and gravel are needed to build a single house, the nearby presence of sand has obvious economic advantages. So, understandably enough, building firms are constantly clamouring for permits from the planning authorities to excavate for sand. But what happens when it *is* excavated? At the bottom of the newly-dug sandpit is the familiar Lower Lias clay, and as water will not drain away through it, the pit soon fills up with rain water. At one time some attempt was made to keep these new ponds as sailing areas for youth clubs but most authorities now tend to insist upon the water being pumped out and the pits filled in with rubbish to make sites for school playing fields or new housing developments. But whatever use they are put to they are certainly going to be a bit of a headache for archaeological societies excavating them again in hundreds of years' time! A pit filled in with the debris of a demolished Regency house plus the refuse of twentieth-century living will be quite a problem to interpret correctly.

THE ORIGIN OF THE SANDS

The grains of the sands are ninety per cent quartz and the remaining minerals indicate that they are derived from Triassic sands from the Midlands. Further evidence of their origin are the facts that the grains are similar in size to desert sand of today, and that if examined under a microscope they have the typical suboval shape of desert sand grains.

As for the method of their journey from the Midlands, this is suggested by their varying thicknesses and by the varying altitudes at which they are found—from about 130 feet to 370 feet near Dowdeswell. These variations indicate a wind-borne action and, as the sands are of recent origin, they could have been carried by winds from the Midlands during a dry glacial period of, say, some 200,000 years ago.

Eroded fossils from the Lias clay are found in some deposits so evidently, in a subsequent wet inter-glacial period, the sands were partially re-sorted by streams running off the Cotswolds. Such wind-blown sands are also to be found in many other parts of the world, including Europe and in vast areas of North China. There they are known as 'loess' and deposits as deep as 1,000 feet have been discovered in North China.

THE ORIGIN OF THE GRAVEL

It will be recalled that in the previous chapter on the Severn terraces presence of gravel was seen to prove that the river had 'been there before', and sometimes large deposits of gravel are found parallel with deposits of the Cheltenham Sands. There are, for example, large gravel deposits at Whitminster, near Beckford, and again at Twyning, where huge quantities are now being

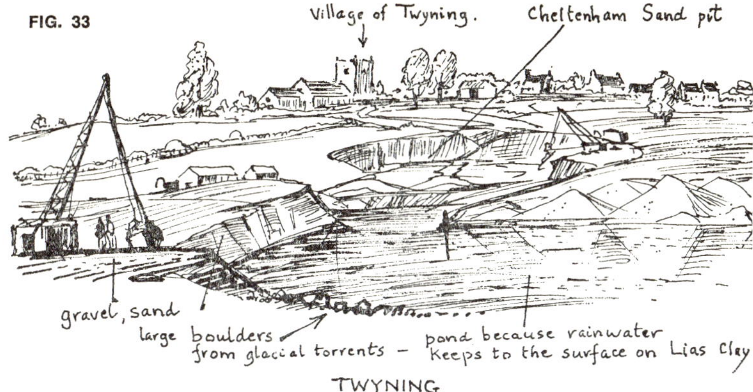

excavated, presenting a dramatic picture of the impact of the later Ice Ages on the face of the earth.

Figure 33 shows that the village of Twyning is sited on the Cheltenham Sands, while deeper down are beds of gravel. At the very base there are quite large boulders of Bunter Sandstone (from the Midlands), Malvern rocks and even large flint nodules. These stones from different areas are a great help to geomorphologists in tracing the movements of the ice-sheets, and even provide a clue to the size and force of the torrents of water which were released when they melted.

The ice lay to the north of the Midlands for many thousands of years and, when it melted, large boulders were swept southwards as far as the Severn Vale. Some of these boulders, called 'erratics', were two feet or more in width and there is one such monster in the Gloucester Museum. Looking at it, one can only marvel at the force and power of a torrent of water capable of transporting such a heavy object for so many miles.

Map 7 and Figure 34 overleaf show the changes which took place after the Ice Ages began some 1½ million years ago, as revealed by the latest 'Carbon 14' dating methods.

Some fascinating hours can be spent in rambling round the excavations at Twyning, looking at the boulders and pebbles of so many different sizes, colours and combinations of mineral content, and determining their different places of origin. Mammoth teeth, bones of deer, and teeth of the woolly rhinoceros have all been found in these gravels, but it is rare for the bones of Palaeolithic

Man to turn up—his remains had little chance of survival in such a hostile environment.

The gravels at Whitminster are not such a collector's paradise. Only a few 'alien' rocks occasionally turn up and most of the gravel consists of Cotswold debris. This is not surprising since, even if these limestone hills were free of ice, the soil itself was permanently frozen and must have generated quite a rapid runoff of water.

THE WATERS OF CHELTENHAM

Cheltonians have more reason than most citizens for taking an interest in the geology of the site of their town, for not only was it originally chosen because of the sands now known as the Cheltenham Sands but its subsequent rapid growth was directly due

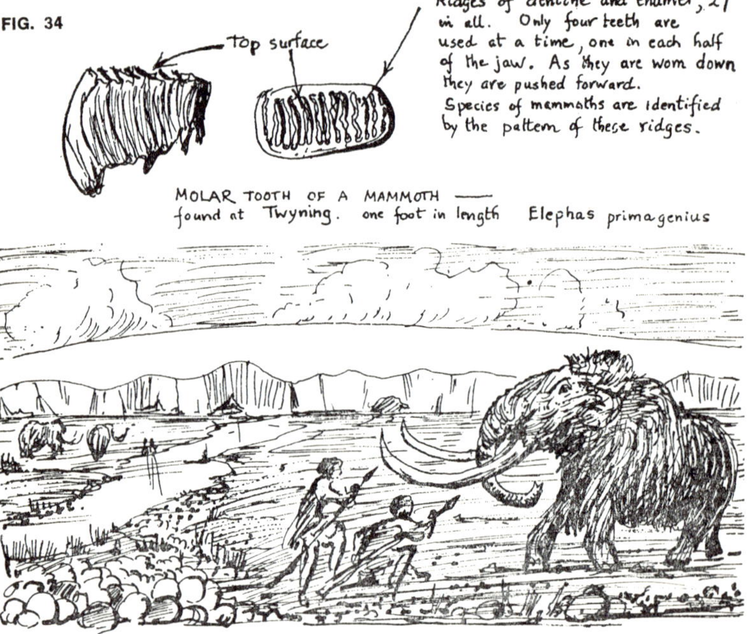

The scene near Bredon Hill by the ice front about 500,000 years ago. Torrents issuing from caves in the ice, a boulder gravel and sand plain Palaeolithic Man of the Acheulian period of culture hunting mammoths,

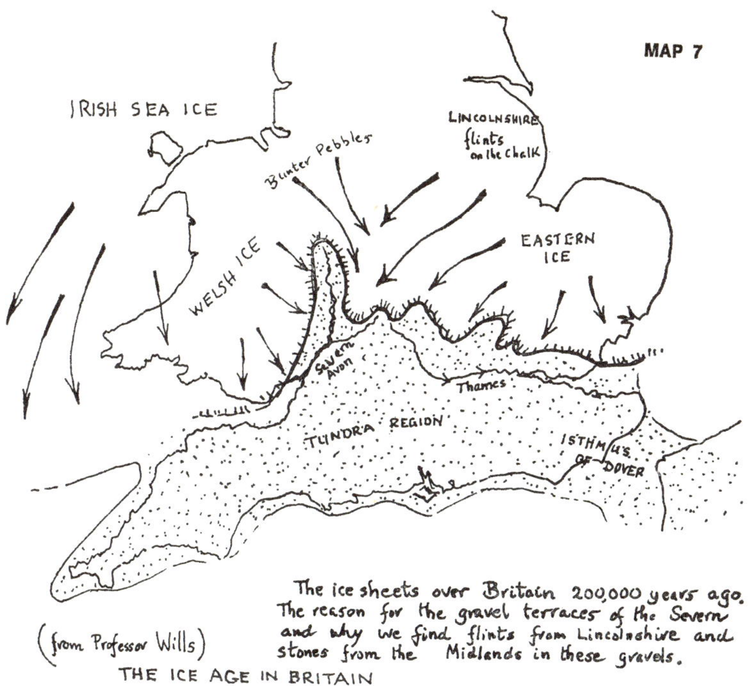

to the nature of the rocks on which it stands.

Wells sunk in the Lower Lias clay almost anywhere in Gloucestershire may yield a kind of saline water and even spring water, issuing from the sand and gravel deposits resting on the clay, is often similarly impregnated. Cheltenham's famous saline waters also originate in the Lower Lias and do not, as was once thought, rise through fissures in the Lower Lias from the underlying Keuper Marl. The problem of their salinity, however, is difficult to understand when one considers the nature of the Lower Lias.

The first spring to be discovered in Cheltenham was found on a Mr Mason's ground and, for about two years after its discovery, it remained open and accessible to people of the town and neighbourhood. In 1718, it was railed in, locked up and a little shed erected over it, its water being sold as a medicine until 1721.

In 1738, Captain Henry Skillicorne not only erected a pump room 'on the west side for the drinkers . . . but protected the spring from

E

all extraneous matter'. About fifty-eight gallons of the water were drawn daily from this, the original Old or Royal Well. But it was George III's visit to Cheltenham in 1788 to take the waters which really put Cheltenham on the national map and new wells were rapidly sunk in many parts of the town. By 1814, as many as thirty-four wells and springs were in use for drinking purposes at the pump rooms or for the manufacture of salts. This number later rose to more than fifty, books and pamphlets proliferated in tribute to the remedial properties of the waters, and the population of the town increased from 1,500 in 1666 to 31,385 in 1841.

Cheltenham is no longer a spa town mainly existing for the exploitation of its waters but even its later growth can be directly

FIG. 35

Diagram to show the origin of the Cheltenham mineral springs. Underground water tends to flow down the direction of dip. The theory that the waters originated from the Keuper Marl is not held today

CHELTENHAM SPA

attributed to its geology. The sudden spectacular popularity of the waters made it possible to plan and build on a grand overall scale never before known in English towns, which had previously grown up piecemeal. Cheltenham is thus a planned town, and one of the first examples of town planning as we know it today.

And how fortunate indeed for Cheltenham that discovery of its mineral waters was made before present-day mass medication! Analyses vary, of course, but most of Cheltenham's famous waters merely contain permutations of those humble ingredients magnesium sulphate, sodium sulphate and sodium chloride! These and other minerals may occur in the waters because of the presence of iron sulphides and bands of limestones in the clays. Percolating water sometimes reaches the surface owing to the general south-easterly dip of the strata (see Figure 35), but all kinds of chemical changes can go on underground with the reaction of the sulphides and the calcium carbonate of the limestones.

Not all the water under Cheltenham, however, is so heavily

impregnated with minerals, and those early Saxons who settled on the Cheltenham sands found good drinking water easy to obtain merely by digging shallow wells in the sand deposits. Today, many of these wells in the sands have been sealed off because they are too liable to contamination from house and garden refuse and might become a danger to health.

When operating under favourable conditions, however, the amount of water which can be gained from these wells is amazing. There are two main wells still working in the Cheltenham Sands, at Sandford Park, by the Lido, and at the Cheltenham Brewery in the High Street. The well at Sandford Park goes twenty-four feet into the sands and yields some 200,000 gallons a day, which is pumped to the outlying hill of Hewlett's Reservoir to get the necessary pressure for distribution. The Cheltenham Brewery well is thirty-five feet deep, with the bottom resting on Lower Lias clay, and the average yield is 27,000 gallons a day.

CHAPTER 8

The Churchdown Outlier

An outlier, as mentioned earlier, is an area of rock separated by erosion from the main mass—and this dissection by erosion is precisely what has happened to the Cotswold escarpment.

It once lay very close to the Severn, and might even have extended as far as the Welsh borderlands, but, in the course of millions of years, the escarpment has been eroded back eastwards —rather unevenly and leaving remnants here and there. However, these outlying hills do have some sort of alignment of their own and Map 8 shows one of the ancient lines of the Cotswold escarpment.

The eroded debris from the escarpment has long since disappeared into the sea, carried there by river water, but these outliers are,

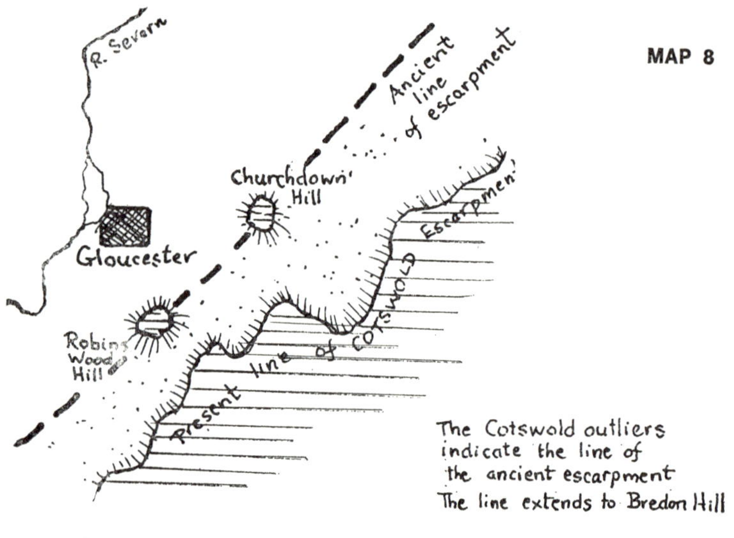

View of Churchdown from Hucclecote

nevertheless, important subjects of study for anyone interested in geology. This is because the Cotswold escarpment as it is today is often covered with slipped-down masses of Upper Lias clay and Inferior Oolite debris which conceals the main outcrops of rock whereas those in the outlying hills are much less obscured. Moreover, being open to inspection 'all the way round', as it were, these outliers are naturally far better places for studying the various divisions of the Lower Jurassic system.

The Churchdown outlier is particularly interesting because it forms a very conspicuous topographical feature of the plain between Gloucester and Cheltenham. Its name, incidentally, is derived from 'Circesdune', a combination of 'Cruc', meaning 'round', and 'Dun', meaning 'hill'—so the name merely means 'round hill' and does not refer to any church in the locality. The general view of Churchdown from any direction is that of a flat-topped or tabular hill and Figure 36 shows it as seen from Hucclecote.

It is best to ascend the hill from the Hucclecote side because changes in its geology are clearly visible in the fields on this side. Round about the 300 ft level, the slope of the ground becomes steeper, coinciding with sandy ferruginous soils in varying shades of rusty brown; a reddish-brown colouring being a marked feature locally of this Middle Lias type of rock. In fact, the main bulk of the Churchdown outlier (half a mile across) consists of Middle Lias rock lying on top of the older Lower Lias.

This is not good arable land. When percolating water reaches the Lower Lias clay below, it creates a slippery underground base and whole masses of the hillside move downhill. The results of this movement can clearly be seen all round the slopes of Churchdown hill, highlighted by the occasional tree which leans drunkenly in an effort to adjust itself to its shifting foundations.

This slip process is typical of all the Middle Lias areas of the Cotswold slopes, so that farmers tend to use such land only as permanent pasture. And even this limited use is often further restricted by prolific invasions of gorse (*Ulex europaeus*) which, as recent research by Dr Eric Jones of Nuffield College has shown, rapidly colonises disturbed ground such as where archaeologists have been at a 'dig', the sites of deserted army camps and airfields or where landslides have occurred. On one side of Churchdown Hill, in particular, where the Upper Lias sands rest on Upper Lias clay, a very considerable expanse of gorse will be noticed.

Further up the hill, on the steep sides, there are numerous badger holes in the sandy soil which these animals prefer, limestone being too hard and clay far too sticky for their claws. These holes are always a boon to the geologist exploring 'in the field' as the badgers' quarrying activities often uncover much interesting detail which would otherwise have become overgrown and passed unnoticed.

At about the 450-foot level, and close to the top of the hill, there is a steep, almost cliff-like, slope caused by a band of hard rock known as the Marlstone, a form of sandy limestone, or ironstone, which often occurs at the top of the Middle Lias and which the amateur geologist should not confuse with the Keuper Marl.

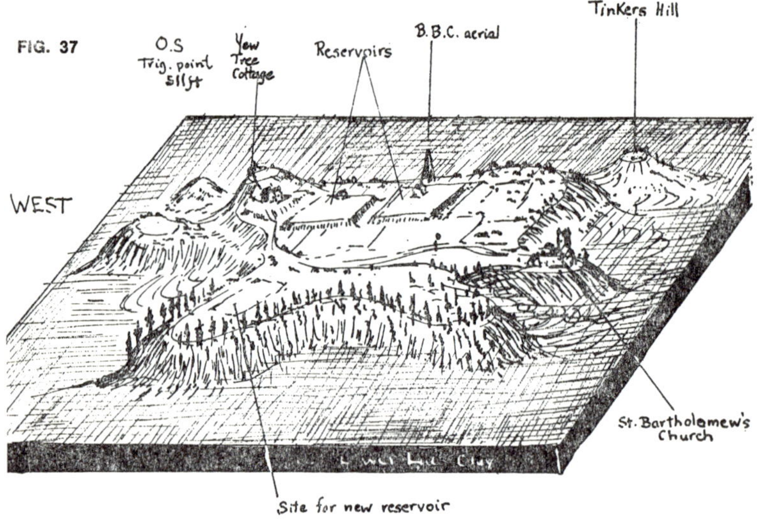

Simplified block diagram of Churchdown Hill outlier. Pine trees grow well on the sandy soils of the Middle Lias

CHURCHDOWN HILL

Although the Marlstone in this case is only some ten to fifteen feet thick, it forms a hard cap to the top of Churchdown Hill and so has saved it from being eroded down to the level of the clay plains (as has happened with the land between Churchdown and the present main Cotswold escarpment). This hard and almost horizontal band of the Marlstone, itself resistant to erosion, is also the cause of the flat shape of the top of the hill.

Figure 37 is a generalised block diagram of Churchdown Hill showing its general form of relief and its main features of general interest.

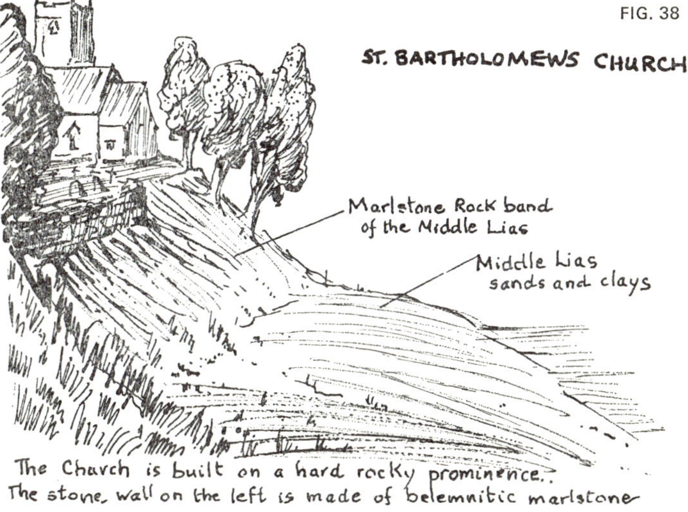

Here at Churchdown, and in this part of the Cotswold escarpment, the Marlstone is much more sandy than elsewhere. Millions of years ago, it was deposited in the sandy bays of the Lower Jurassic sea and as many shell-bearing animals lived in those shallow seas the Marlstone rock is very fossiliferous, the most noticeable fossils being bivalves (e.g. the Pectinids—*Pecten*) and belemnites. It is also a very ferruginous type of rock, which is why springs issuing from the Middle Lias are often a rusty brown in colour and have a high iron content.

The church of St Bartholomew, on Churchdown Hill, is built on a rocky eminence of the Marlstone, but though the stone was once quarried there—remains of the old quarry lie between the church and the reservoirs—it is not good building material because the fossils are large and, when attacked and etched out by frost, give the rock a ragged look. This is particularly noticeable on the wall by the church where the dominant fossils in the Marlstone blocks can be plainly seen.

The Middle Lias Marlstone has a general tendency to form plat-forms with flat summits and steep sides because the soft sands and clays of the Upper Lias, which lie above this rock, are as rapidly eroded as the soft clay of the Lower Lias which lies below it. This erosion battle can be observed to particularly good effect on Churchdown Hill, the highest point of which is on the western corner, by Yew Tree Cottage, where a tenacious remnant of Upper Lias sands reaches a summit of 511 feet. Figure 39 explains this.

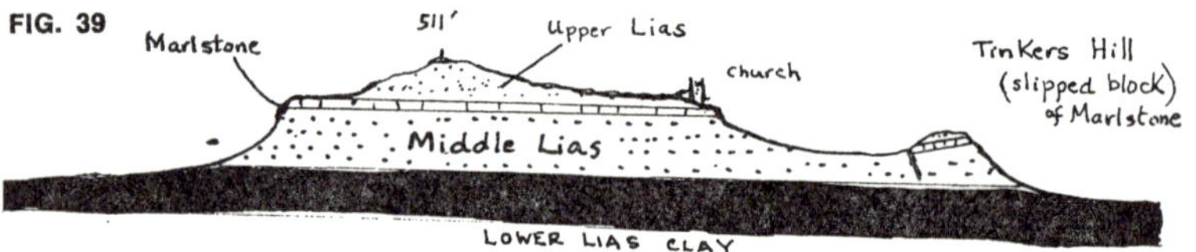

Geological section of Churchdown Hill   The Middle Lias is about 150 ft thick

As for the adjoining Tinkers' Hill this, in the writer's view, is a slipped-down mass of Marlstone, as it reaches a height of only 350 feet, whereas on Churchdown Hill the Marlstone outcrops at about 450 feet.

The hill already has large covered-in reservoirs on its summit and more are to be added in the near future to cater for the ever-expanding populations of Cheltenham and Gloucester. The hill is also the site for a BBC booster aerial for television—an equally vital modern commodity.

But the 'folk who live on the hill' probably do not worry over-much about progress on these reservoirs. For surface rainwater on the hill is retained in the water-holding formation of the Marlstone —known as an 'aquifer'—and sinks slowly through the sandy beds of the Middle Lias to be released at springs where the junction occurs with the impervious Lower Lias clay. A good example of one of these springs can be seen in use by the farm to the north of The Green at Churchdown.

CHAPTER 9

Robin's Wood Hill

Robin's Wood Hill, only two miles from the centre of Gloucester, was once known as 'Mattes Knowle'—the name, like that of nearby Matson, deriving from the de Matteson family who became lords of the manor after the Norman Conquest. Matson still perpetuates their memory but the hill now bears the name of an old Gloucestershire family, the Robins, generations of whom leased the land for sheep-farming from 1526 until 1759.

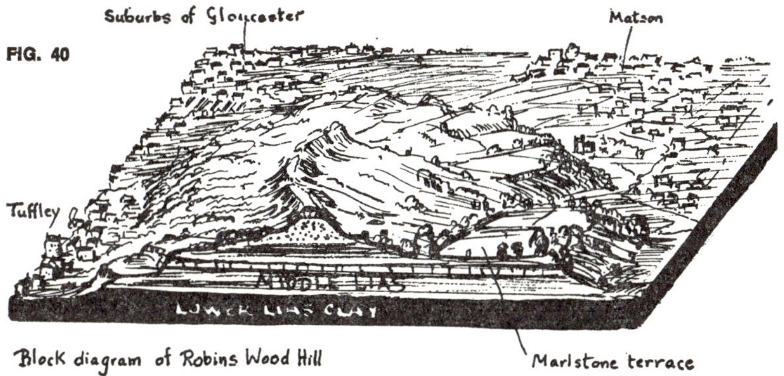

Block diagram of Robins Wood Hill   Marlstone terrace

Robin's Wood Hill has known other occupations beside farming, however, and at its foot there is still to be seen an old brickworks, now derelict, which was once the prosperous and well-known Tuffley Brickworks (Figure 41). The bricks were made from the Lower Lias clay which forms the foundation of the hill and, in its natural state, is bluey-grey in colour owing to the presence of finely-disseminated iron sulphide. On exposure to weather, hydration of the iron minerals changes the colour of the clay to brown. But the real misfortune of the Tuffley Brickworks, as of other once equally-famous but now abandoned local brickworks, was that, by comparison with an Upper Jurassic clay, Lower Lias clay is an unreliable material for modern brickmaking and unless production

FIG. 41

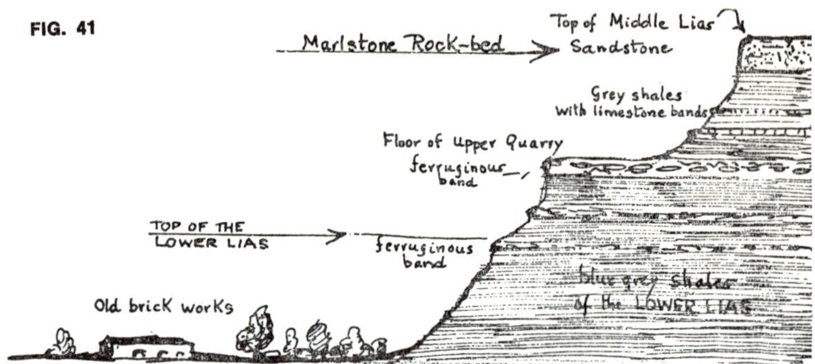

THE STRATA AT TUFFLEY OLD BRICKWORKS, ROBINS WOOD HILL

is scientifically, and expensively, controlled bricks made from it will often shatter during firing or else disintegrate later on in the walls into which they are built. Today a greater part of the 760 million bricks produced each month in Great Britain are made from an Upper Jurassic clay—the Oxford Clay near Peterborough.

At the abandoned Tuffley Brickworks relics of the chancy business of using Lower Lias clay may still be seen—blue-grey bricks left where they were being air-dried, red bricks which look perfectly sound, and fragments of shattered red brick which have on them a white crust of sodium sulphate.

There should also be evidence nearby of another abandoned industry—an ironworks, but all trace of it now seems to have vanished. Yet the records of its existence are quite clear and a Government publication of 1952, 'The Liassic Ironstones' (Geological Survey Memoirs), stated:

'According to P. B. Brodie (in Gavey 1853, p.31) ironstone was formerly worked at Robin's Wood Hill, near Gloucester, and Woodward (1893, p.31) adds that the forging of iron appeared to have been carried out there to a considerable extent, the ore being dug out of the hill.'

Any geologist looking up at the cliffs of Tuffley quarry will find this quite credible for, higher up, the cliffs begin to change colour. The blue-grey of the Lower Lias clay rock-face gets more and more sandy until a thin band of ironstone nodules can be seen, their number steadily increasing as the eye travels higher up the rock-face. In fact, the iron content is now only about ten to twenty per cent, which is not high enough to make it worth mining.

These are the same ferruginous rocks as those in the Middle Lias of Churchdown Hill and, when traced northwards, can be seen to have given rise to the early steel industries of Britain. It appears at Cleveland in Yorkshire, in the great open-cast quarries of Grantham, Melton Mowbray and in the Banbury districts. Today, however, more than half of the total output of British iron-ore comes from the Inferior Oolite, the rocks of the Middle Jurassic series. This is called the Northampton ironstone and extends from Lincoln to Grantham.

THE CLIFF AT TUFFLEY

At the top of Tuffley quarry there is another cliff-face set further back, and this 'near top' of the Lower Lias rocks forms a convenient grassy platform on which to walk round and look up at the finest inland exposure of the Middle Lias rocks to be found in England. Here grey shales finally culminate at the top with the Marlstone

FIG. 42

THE MARLSTONE ROCK-BED ON ROBINS WOOD HILL

rock-bed, which is here about fifteen feet thick. It is a buff-coloured, fine-grained sandstone, which glows with colour in the sunlight—a bold and handsome cliff, not dingy and crumbling like the clay cliff below.

Look at it closely and you will see why it gleams in the sun. There are fine flakes of mica in the sandstone, and if these are examined through a hand lens it will be noted that the tiny grains of quartz sands are coated with a rusty, powder-like mineral. This is limonite, and the cause of the rock's colour.

Looking south from here to beyond Stroud and Nailsworth, this same rock-bed forms the Cotswold sub-edge where the under-edge or shelf is a few miles wide and the site of several village settlements. The ground hereabouts is strewn with great blocks of stone which have tumbled down from the Marlstone rock-bed on to the grassy platform, so that the whole place has a kind of Easter Island atmosphere. Most of the rocks are rectangular, as that is how the Marlstone has fractured, but there are also a number of huge spherical boulders which were formed by concretions on the sea floor during the Middle Lias period or perhaps by compaction of the rocks later during processes of cementation. These 'doggers', as they are called, erode out of the main mass of rock and look rather like huge footballs.

THE EVOLUTION OF A MINERAL

The most interesting feature of this rock, however, is its iron content, derived from two minerals—siderite (iron carbonate) and chamosite (hydrated iron aluminium silicate). These minerals are not pure and it is believed that the first to be formed was chamosite, which later altered to siderite which, in turn, became limonite after weathering. So, today, when we look at a russet-brown nodule of limonite we are actually seeing two changes brought about by weathering—not only the wearing away of the rock but an actual change in its mineral content. We are, in fact, looking at the end-product of a most complicated process in the evolution of a mineral.

This still does not explain how the siderite and the chamosite came to be in these marine rocks in the first place. We do not yet know for sure but it has recently been suggested that their presence is due to the existence of 'iron fixing' bacteria in those ancient seas. It is only lately that geologists have realised the geo-chemical importance of bacteria, and have been able to observe the results of

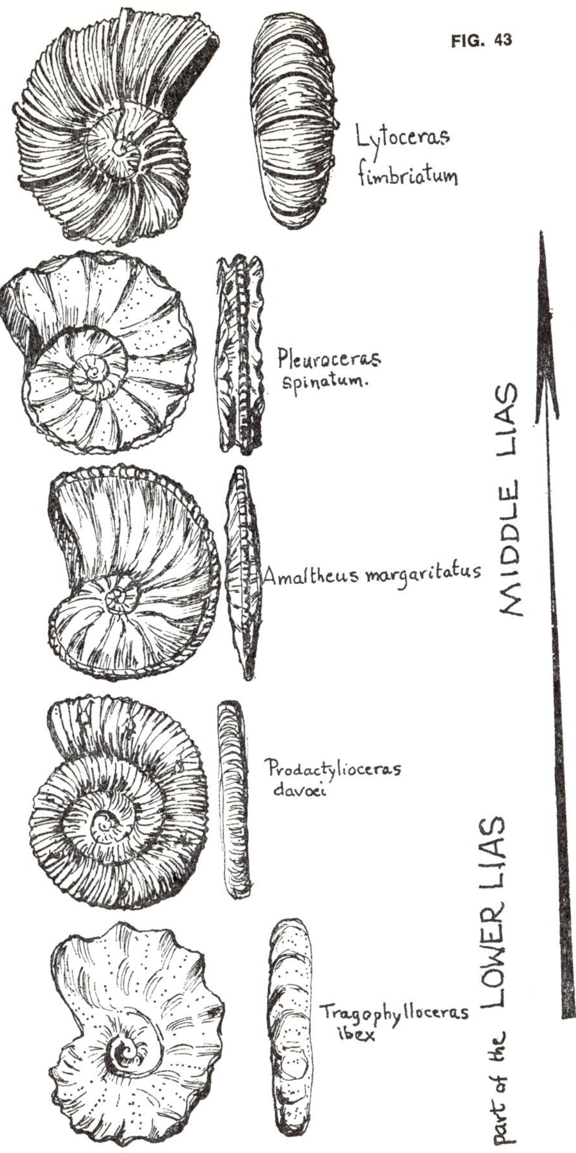

Ammonites used for identifying various divisions of the top of the Lower Lias and the Middle Lias. Several million years separate each species

millions of bacteria depositing iron ore in the tropical seas of today. Both bacteria and algae must have played an enormous part in the creation of chemical deposits but, other than this instance, they have left no trace of their existence.

'THE AGE OF THE AMMONITES'

The rocks of Robin's Wood Hill were laid down during the Jurassic period, which is famous among geologists as 'the age of the ammonites'. Ammonites became extinct some 70,000,000 years ago but they evolved so rapidly during the Jurassic period that the various layers of the Lower and Middle Jurassic rocks can be identified by the type of ammonite found in them.

Figures 43 and 44 shows various types of ammonites. Notice how they differ from one another in the type of coil and ornamentation.

Ammonite discoveries have been useful in 'zoning' the types of rocks on Robin's Wood Hill but they have been more commonly found in the clays at the base of the hill rather than in the sandy beds, where they tend to get dissolved by percolating water to such an extent that only the casts or moulds are left. The ammonite casts are of shells which covered the body of an animal related to the present-day nautilus and which looked something like an octopus with a shell round it.

Ammonites are now difficult to find on Robin's Wood Hill but

FIG. 44

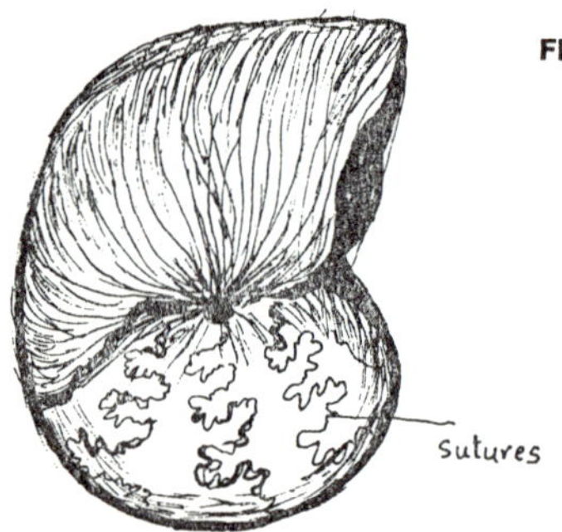

Phylloceras heterophyllum.
An AMMONITE from the Upper Lias.
Part of shell removed to show sutures

Pseudopecten equivalvis.
A bivalve (lamellibranch)
from the Middle Lias.

α PECTEN ÆQUIVALVIS

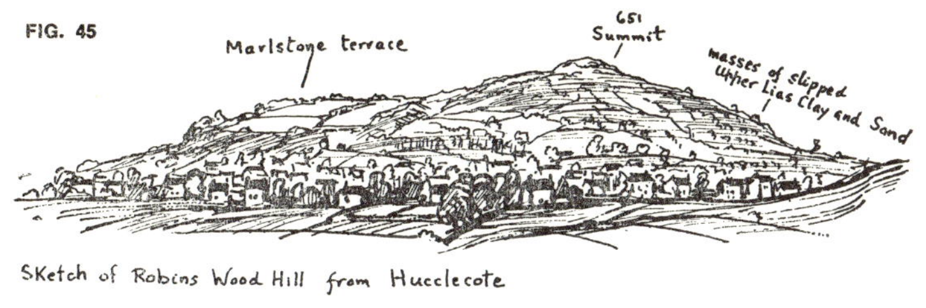

Two sketches which explain the shape of Robins Wood Hill
The Marlstone Terrace is well developed only on the eastern side

one piece of geological evidence, still apparent despite the thick
pasture covering it, is the Marlstone rock-bed which forms a few
remarkably flat fields on one side of the hill. These are so level as
to look almost like a man-made sports ground and are, of course,
repeating the tabular shape of the Churchdown outlier where the
Marlstone occurs on that particular hill.

The rest of Robin's Wood Hill, by contrast, is very irregular in
shape, having a partial covering of Upper Lias clays and sands plus
a small capping of Inferior Oolite, the same kind of limestone which
caps most of the Cotswold escarpment.

CHAPTER 10

Bredon Hill

Bredon Hill is the largest Cotswold outlier, a whale-back of a hill lying athwart the southern Midlands in Worcestershire but also right on the northern Gloucestershire boundary line. It is about three and a half miles long by one and a half wide and deflects the north/south traffic, which has to go through Tewkesbury in Gloucestershire or via Evesham in Worcestershire. Some 20,000 years ago, in an early glacial period, it was a barrier to a different kind of movement—a converging ground for ice masses coming down from Wales and Lincolnshire.

The absence of any main road over the hill, plus the fortunate accident of single ownership, makes it possible today to walk its entire length through surroundings of incredibly unspoilt beauty. From its highest points there are panoramic views which must be some of the loveliest in England—particularly those as one looks west towards the mountains of Wales.

But the geologist, when he has looked his fill, will eventually remind himself that the views are just as much the product of geological trends as the hill from which he sees them and which he is now studying. Actually, he will have begun his study of Bredon Hill before ever he sets foot on it as its shape, seen from a distance, at once betrays the secrets of its geology. Particularly good 'end on' views of the hill can be obtained from several places near Upton-on-Severn. Figure 46 is such a view and shows how very obviously the strata of the hill dip southwards.

FIG. 46

Bredon Hill — an 'end-on' view from near Upton on Severn.

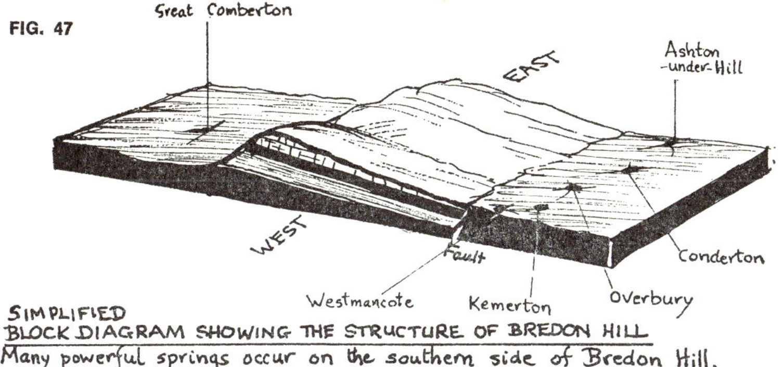

SIMPLIFIED
BLOCK DIAGRAM SHOWING THE STRUCTURE OF BREDON HILL
Many powerful springs occur on the southern side of Bredon Hill.

Figure 47 shows how the slope of the ground is controlled by the structure of the rocks on Bredon Hill. Figure 48 is a view from one of the many agreeable walks from the summit down the dip slope to Overbury. Figure 49 overleaf depicts the fine escarpment which can be seen on the northern edge of the hill. With a dip slope to the south, the geologist would expect to find just such a dramatic escarpment on the northern edge and here it is displayed to perfection.

OOLITE LIMESTONE

This is limestone country, and Bredon Hill is capped with Inferior Oolite limestone, a Middle Jurassic period rock whose peculiar texture distinguishes it from some other limestones laid down in a sea.

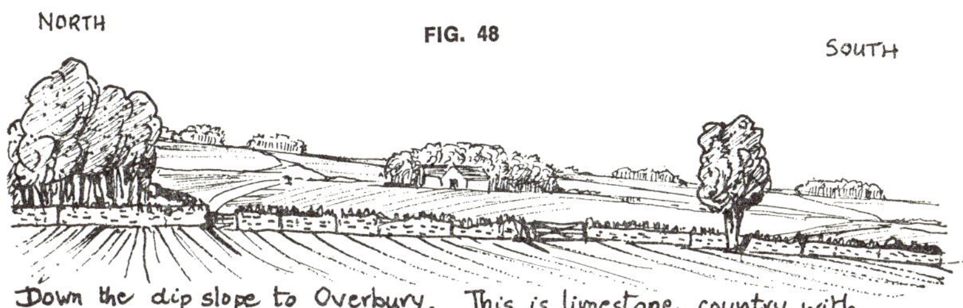

Down the dip slope to Overbury. This is limestone country with stone walls and large fields. Some are as large as 75 acres. The beech tree clumps are used as pheasant covers.

DIP SLOPE SCENERY ON BREDON HILL. Topography controlled by structure.

F

Though technically known as 'Oolite', meaning 'egg stone', quarrymen have a simpler and far better name for this rock. They call it 'roe-stone', because it resembles herring-roe.

An oolite is made up of numerous small spherical bodies called 'ooliths', usually less than a millimetre in diameter and cemented together by calcite. Under a microscope, the ooliths are seen to be made up of concentric layers around a nucleus, which may consist of a shell fragment or a grain of quartz.

Oolites occur in many limestones of different ages and they can even form under artificial conditions in a factory boiler if the water is hard, i.e. saturated with calcium bicarbonate, and if the water is kept vigorously agitated. In the shallow coral seas between Florida and the Bahamas, in the Red Sea and Persian Gulf, ooliths are constantly being formed and it follows that oolitic limestones will

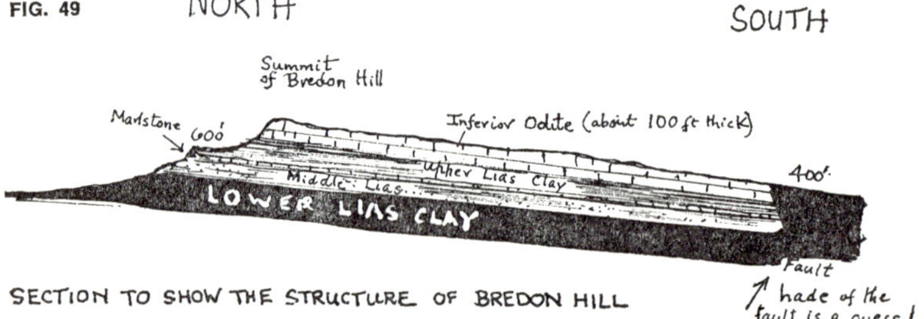

be formed in seas saturated with calcium carbonate and where tiny grains on the sea floor are kept constantly in motion by currents. As the grains drift backwards and forwards they become covered with concentric layers of calcium carbonate.

Such a fine-textured limestone will form a good building stone —particularly if all the fossils in it are broken up into tiny fragments. Bath stone, which is called the Great Oolite, is a stone like this and has been quarried for building for many hundreds of years.

Underneath Bath stone there is another kind of oolite, not so good for building, for which William Smith suggested the simple name 'Under Oolite', but whose official name is 'Inferior Oolite'. Although this is less than thirty feet thick in the Bath region, it becomes over 200 feet thick when traced to the North Cotswolds and forms the main building stone in that region. The Inferior Oolite on the North Cotswolds is thicker because sedimentation covers a longer period of time with a slowly sinking sea floor.

Recent research has also shown that some oolitic limestones have been formed organically from myriads of calcareous algae.

THE SCENERY OF BREDON HILL

To make a detailed geological interpretation of the scenery of Bredon Hill it is best to walk along the escarpment extending east and west of Bredon Tower, the small, rather rugged 'summer house' built round about the year 1800 on the highest point of the hill, and then walk down the long dip slope to Overbury.

Diligent readers of the one-inch-to-the-mile Ordnance Survey map of this area will notice that it does not give the height of this highest point. Estimates which the writer has collected have varied from 'only a few feet short of 1,000 ft' (this from the estate manager and from patrons of the inn at Overbury) and 977 ft (this being the estimate of L. E. Richardson, the famous geologist who has made special studies of both the Severn Vale and the Cotswolds).

Down the dip slope will be seen some of the largest fields in the Gloucestershire/Worcestershire area, one of these being 75 acres in extent. Notice that the soils derived from Inferior Oolite are russet-brown in colour, and that the fields are strewn with fragments of the limestone, some large enough to damage the plough —a constant hazard this of ploughing on the Oolite.

The russet colour of the soil is due to the limestones themselves being coloured by the presence of iron oxides. The iron in the unweathered rock is probably in the form of iron carbonate, which is pale in colour. When the rock weathers, the carbonate is decomposed and the iron is left in the soil as iron hydrates and oxides which, on complete oxidation, are red and brown. As the limestones are dissolved away by rainfall and percolating water, only hydrated iron salts (mainly limonite) remain. This is exceptionally well seen in Mediterranean countries where vast areas of limestone rocks are covered with a thin reddish soil called 'terra rossa'.

Notice that barley is the main crop here. Cotswold farmers have discovered that it stands up to drought conditions far better than wheat, and so makes a good crop for limestone country where there is rarely any surface water, limestone being a rock through which water drains away into the fissures.

There are many public footpaths leading down the dip slope to the strings of villages which fringe the southern side of Bredon Hill and these give ample opportunity to observe how everywhere there

FIG. 50

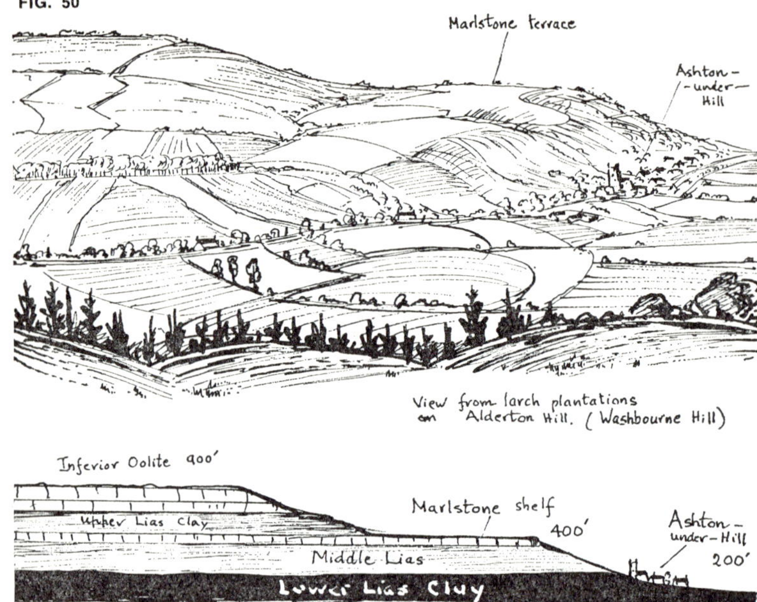

The Marlstone platform is well developed only on the eastern side of Bredon Hill, in the same way as Robins Wood Hill.
BREDON HILL — EASTERN END

are marked correlations between the structure and type of rock and what is being grown.

The limestone country rolls down to Overbury, its way marked by dry valleys, clumps of beech trees and many old quarries. Coming to the southern edge of the hill, a huge fault can be seen to have so displaced the rocks that the Inferior Oolite has sunk to the level of the much older Lower Lias clay, a soil particularly attractive to elm trees because of the secure anchorage it affords to their shallow roots.

A glance back at Figures 47 and 49 will show the exact relationship of these totally different strata and a careful study of the country between Westmancote and Ashton-under-Hill will reveal on one side the rich lush grass of the clay and, adjoining it, the rolling topography of the limestone.

SUCCESSION OF STRATA ON BREDON HILL

The Inferior Oolite is about 100 feet thick and begins to outcrop at about 870 feet on the north scarp, whereas on the south it is brought down by the fault and dip to 400 feet. Immediately below the limestones (the rather hummocky ground and slipped-down masses of oolite) are the surface indications of the Upper Lias clay.

The Marlstone of the Middle Lias outcrops at 600 feet near Woollas Hall, but is found at 400 feet near Bredon's Norton, which means a drop of 200 feet due to the fault. As on the eastern side of Robin's Wood Hill, there is also a very well-developed Marlstone shelf on the eastern side of Bredon Hill—yet in other parts of the hill there are only small remnant shelves of Marlstone. The writer believes that this differential erosion may be the result of Pleistocene (glacial and inter-glacial periods) erosion.

THE KING AND QUEEN AND THE BANBURY STONES

If a relatively hard rock like the Oolite rests on a weak foundation rock (e.g. the Upper Lias clay), it will tend to slide downhill on it, and this happens very often in the Cotswolds. On Bredon Hill, in particular, this tendency is greatly accentuated by the fact that the limestones were originally horizontal but are now tilted southwards and downwards towards the fault line. This causes the vertical joints, which are at right angles to the bedding planes, to open out until they eventually become filled with masses of broken-up Oolite and the whole lot is cemented together by percolating water rich in carbonate of lime. This petrifying action produces a natural concrete.

The fissures thus filled up are called 'gulls' and while some fine examples—the Banbury Stones—are to be seen on the summit at Bredon Hill Tower, there is a far better group, known as The King and Queen Stones, near an old quarry just above Westmancote.

It seems probable that, in these particular cases, the gulls have become harder than the surrounding rocks, which have gradually been eroded away, leaving the gulls standing as large upright slabs.

Some highly imaginative stories have grown up around these stones, and local opinion strongly favours the Druids as having been responsible for their erection. This is most improbable, but the Banbury Stones are within the 'precincts' of a very fine Iron Age

FIG. 51

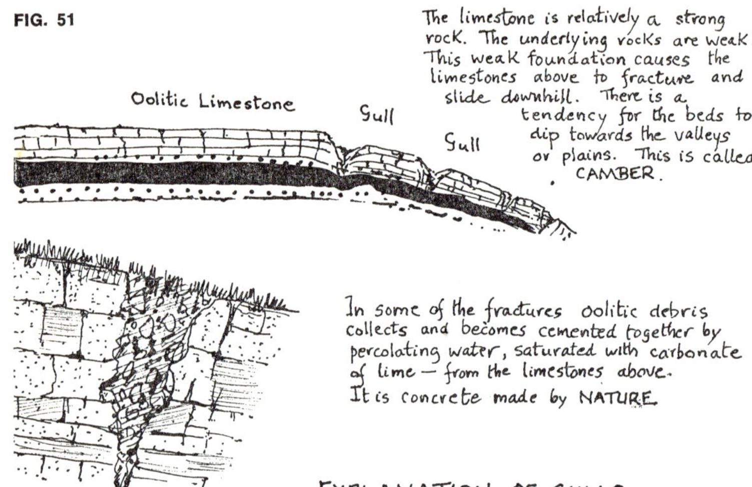

EXPLANATION OF GULLS.

GULLS AND THE KING AND QUEEN ROCKS

earthwork and it may well have been that the inhabitants of this fort used the slabs as convenient places for assembly.

After studying the King and Queen Stones, the opportunity should be taken to examine the disused quarries nearby. Between five and ten feet down from the top of each quarry can be seen curious circular patterns of rock fragments, probably caused by convectional rotary movements of rock during the freeze-thaw conditions of the last Ice Age.

FIG. 53

These curious structures can be seen in many Cotswold quarries, usually about 5 or 10 feet down. They may be due to convectional rotary movements of rock fragments during freeze—thaw conditions of the last Ice age. Perhaps they may be due to ice lenses which form on the subsoil under intensive frosts.

Structures observed in a quarry above Westmancote, near the King and Queen rocks.

SPRINGS AND VILLAGES ON BREDON HILL

Much of the rainfall on the hill sinks through the fissures in the limestone to reach the Upper Lias clay and, being unable to penetrate the clay, it then travels southwards along its surface eventually to be released at numerous springs along the southern edge of Bredon Hill. Obviously, the villages along there owe their origin to the existence of these never-failing water resources.

A good place to see the springs in action is in Overbury Park because here the Upper Lias clay floors the valley. Many springs also issue from the Middle Lias either just below the Marlstone rock-bed or at the junction with the sandy beds of the Middle Lias (here about 200 feet thick) and the Upper Lias clay. Such springs can be seen near Elmley Castle and at Ashton-under-Hill.

CHAPTER 11

Leckhampton Hill

Queen of the deserted quarries of the North Cotswolds is undoubtedly that at Leckhampton, two miles south of Cheltenham and now owned by Cheltenham Borough Council, who acquired Leckhampton Hill as a social amenity for the city in 1928.

Officially opened in 1793 and the source of much of the stone used in the building of Regency Cheltenham—in 1810 blocks of dressed stone cost only 1d per ton delivered!—Leckhampton Quarry today attracts student geologists from all over the country, many of whom travel long distances at week-ends to see the dramatic exposures it offers of what might be described as 'the innards' of the Cotswolds.

Approaching Leckhampton Quarry from Cheltenham, the first steep part of the road is where the Marlstone shelf of the Middle Lias begins. A large building, Hill House, is perched snugly on this

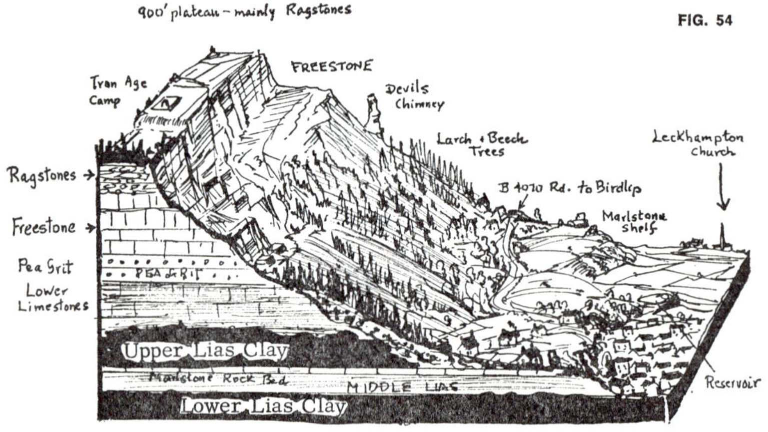

BLOCK DIAGRAM OF LECKHAMPTON QUARRY, looking west

shelf and the road takes advantage of a small gap in the rock-bed to climb up to the Cotswolds.

The official entrance to the quarry is at the car park in Daisy Bank Road and here the steep slopes on the lower sandy beds of the Upper Lias are planted with larch trees to stabilise the soil. Here, too, the ground is everywhere hummocky, the result of the slipping downhill of the Upper Lias sands and blocks of limestone over the Upper Lias clay which acts as a lubricated surface.

From this point the course of an old railway line can be followed to the ruins of the old lime-kilns which were in use as late as 1927 (see Figure 56). From the kilns, the limestone cliffs rise to impressive heights, attaining 965 feet at their highest point. From top to base there is a grading of colour from cream to russet brown, indicating that the rocks become more ferruginous towards the base of the cliff.

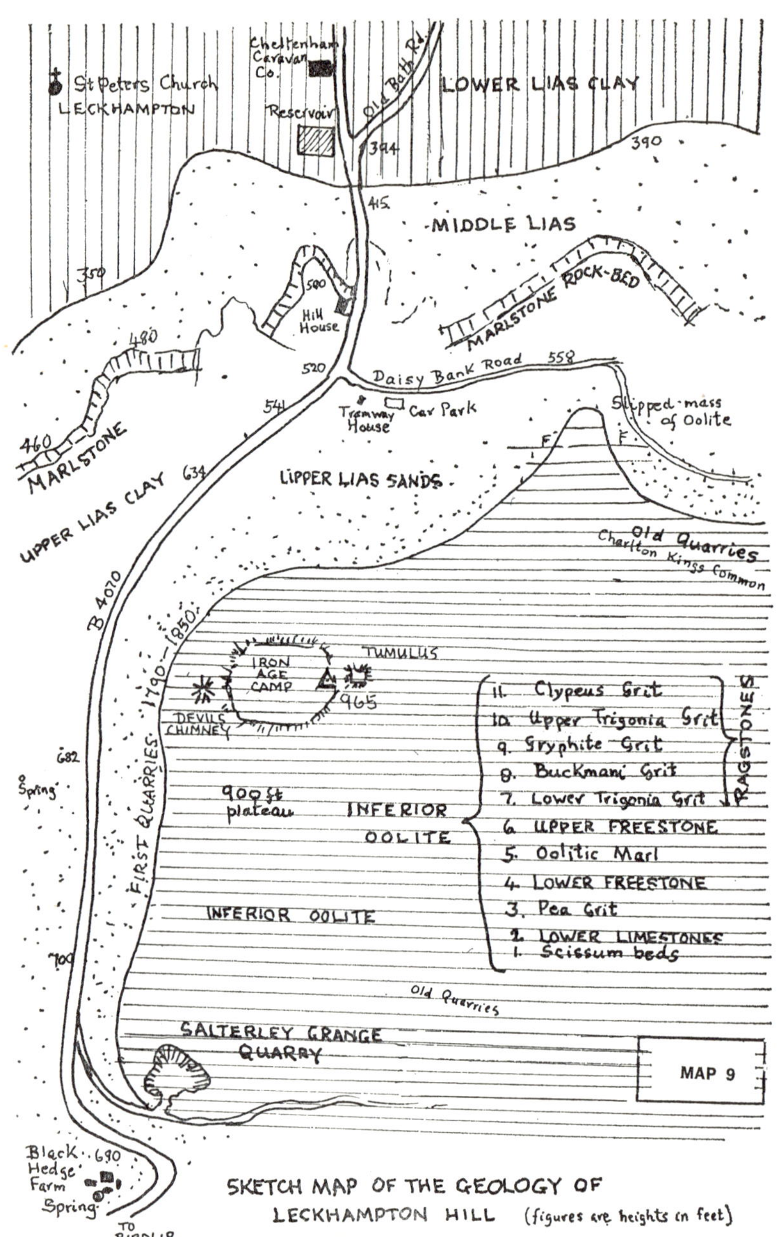

St Peters Church
LECKHAMPTON
Cheltenham Caravan Co.
Reservoir
Old Bath Rd.
LOWER LIAS CLAY
394
390
415
MIDDLE LIAS
350
MARLSTONE ROCK-BED
500
Hill House
480
520
Daisy Bank Road
558
541
Tramway House
Car Park
Slipped mass of Oolite
460
MARLSTONE
F
F
UPPER LIAS CLAY
634
UPPER LIAS SANDS
Old Quarries
Charlton Kings Common
B 4070
First Quarries 1790-1850
IRON AGE CAMP
TUMULUS
965
DEVILS CHIMNEY
682
Spring
900 ft plateau
INFERIOR OOLITE
11. Clypeus Grit
10. Upper Trigonia Grit
9. Gryphite Grit
8. Buckmani Grit
7. Lower Trigonia Grit
6. UPPER FREESTONE
5. Oolitic Marl
4. LOWER FREESTONE
3. Pea Grit
2. LOWER LIMESTONES
1. Scissum beds
RAGSTONES
INFERIOR OOLITE
700
Old Quarries
SALTERLEY GRANGE QUARRY
MAP 9
Black Hedge Farm
690
Spring
TO BIRDLIP
SKETCH MAP OF THE GEOLOGY OF
LECKHAMPTON HILL   (figures are heights in feet)

THE PEA GRIT

Just behind the old kilns are the rather crumbly cliffs made of Pea Grit, a rock composed of ovoidal bodies about the size of a pea. This structure is known as 'pisolitic', and the pisoliths are similar to the egg-shaped ooliths in the Oolitic limestone above but larger. Their origin is not fully understood, but one theory is that they have been formed by vigorous agitation of nuclei (small grains) in a sea which was actively precipitating calcium carbonate. 'Cave pearls' are believed to have been formed in this way.

An alternative theory is that the pisoliths have been formed by calcareous algae around a nucleus. The name *Girvanella pisolitica* has been applied to this type of algae and recently it has been discovered that if pisoliths are treated with acid, the algal filaments can be seen still preserved.

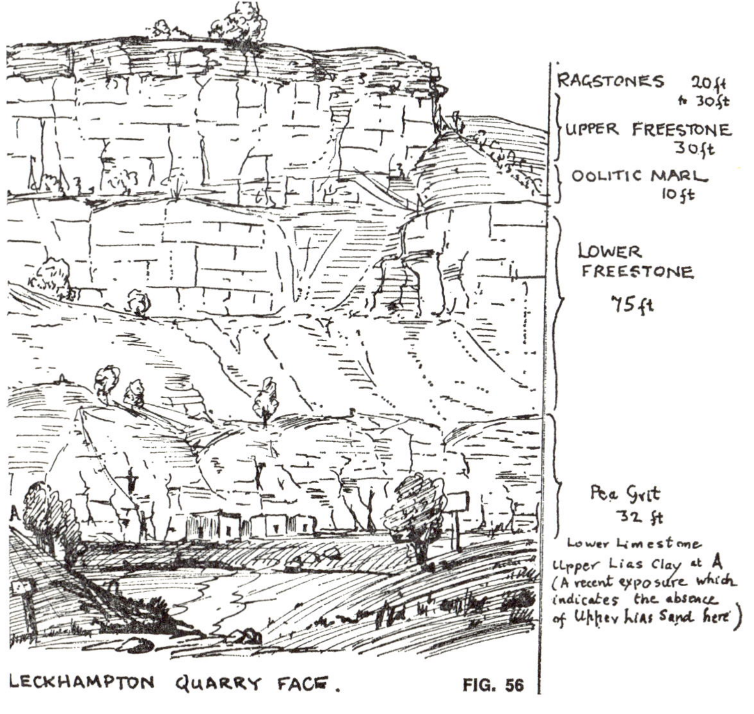

LECKHAMPTON QUARRY FACE.   FIG. 56

Fragments of crinoids, corals, echinoids and brachiopods can be found in this Pea Grit, which is here about thirty-two feet thick and can be traced all along the hill low down, below the Devil's Chimney and away over to Crickley Hill.

It is a limestone which has been formed in a sea in which calcium carbonate was being precipitated, and iron in the sea water was probably being precipitated by bacteria. But it is not a good building stone—frost action can make it disintegrate, the pisoliths falling apart from each other.

THE FREESTONE

The limestone cliffs above the Pea Grit belong to the Inferior Oolite division of the Middle Jurassic system. Their most striking feature is the rather regular spacing of the horizontal bedding planes and the joints at right angles to them, giving a general appearance of man-made walls. In fact, this kind of rock is described by some geologists as 'mural jointing'.

A bedding plane is produced by a pause in the process of sedimentation on a sea floor. It can also occur if there is a change in the type of sediment being laid down. Where there is a thick mass of rock, sedimentation has obviously been continuous for a long time.

When the bedding planes are far apart the rock is called Freestone because neatly rectangular slabs can be removed 'freely' from the quarry face, and these make a good building stone for the type of classical architecture found in Regency Cheltenham. Another school of thought, however, holds that the term 'Freestone' derives from the fact that this particular stone is so devoid of large fossils and of such an even, fine-grained texture that it can be *freely* sawn into blocks.

If the bedding planes are close together, creating thin wedges of rock, the rocks are called Ragstones and these can be seen right at the top of the quarry.

More often, however, the Ragstones are bedded hard limestones (often ferruginous) which break up irregularly and are well stocked with fossils. In the past they were extensively quarried for local road metal and at the top of the quarry there are many hollows and banks indicating where quarrying took place long ago.

Hundreds of people come to Leckhampton Quarry equipped with hammers ready to knock out fossils in the Freestone—and they are

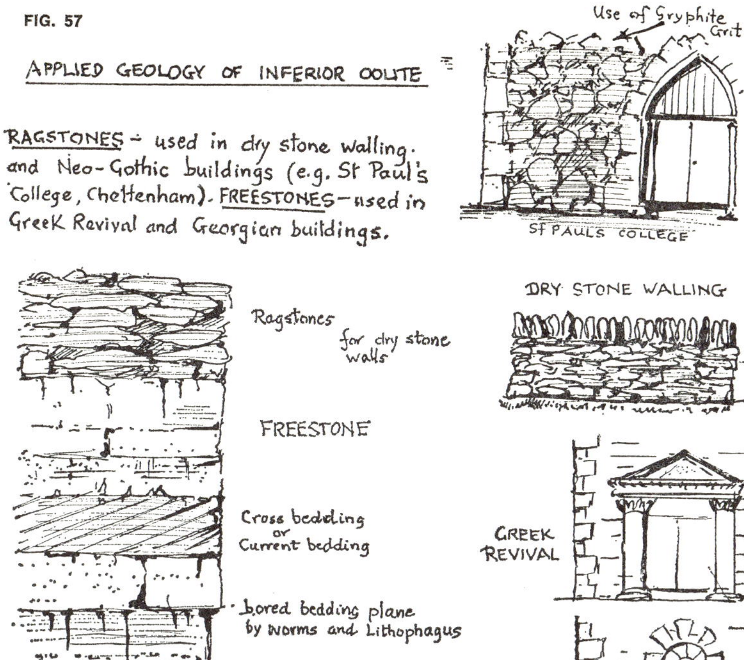

disappointed! A close look at the rock will reveal fossil remains but they are merely thin bands of minute fossil fragments which were broken up by the sea floor currents at the time of deposition. This is another factor which makes the Freestone a good building stone— the fossils are so finely comminuted that there are no large fossil remains for frosts to ease out and thereby pockmark a wall with holes.

CURRENT BEDDING IN THE FREESTONE

A change in the velocity of the currents in a sea will cause changes in the nature of deposition. This is called 'current' bedding, or 'cross' bedding, and it can be seen here and there in many parts of the quarry, particularly in the Freestone. The occasional slab of stone which is produced under such conditions is unpopular with builders—they know it will develop a ragged appearance as it weathers and will sometimes produce a damp spot on a wall.

When limestones are first deposited they are, of course, soft and during the process of compaction tensions are set up in the rocks as the rocks dry out. In addition to this stress, the calcium carbonate often crystallises into calcite, which causes the rock to expand.

The stresses set up in the rocks tend to pull them apart, so tension joints appear at right angles to the bedding planes. If the rocks are very much the same all over (i.e. 'homogeneous') then the joints are well developed and regular, which makes it easy to get the rock out of the quarry in convenient shapes ready for building use.

These rocks are also subjected to other stresses, e.g. compression, and this results in sheer joints which penetrate all the bedding planes at right angles. Some of the forces involved are intense and operate over wide areas, even to the extent of mountain building.

Just near the bedding planes many of the limestones are riddled with small holes—especially at the top of cross-bedding structures. This probably means that soon after the rock was formed worms bored into the rock, though the borings could also have been caused by a bivalve creature called *Lithophagus*. This mollusc, which lived 150 million years ago, had an exterior shell shaped like a rasping file which enabled it to bore into rock. It has also recently been discovered that many borings could have been done by a group of organisms known as Phoronids—worm-like animals which live in tubes.

THE DEVIL'S CHIMNEY

This is a column of rock jutting out from the face of Leckhampton Quarry. It was a famous landmark even in the early nineteenth century, being first mentioned by Ruff in his *History of Cheltenham* (1803), when he wrote: 'Built by the devil, as say the vulgar. It was no doubt built by shepherds in the frolic of an idle hour.'

As this part of the quarry was actively worked about 1780 it is more likely that quarrymen removed the surrounding stone but left this particular column of rock because it was not good enough to be used as building stone. A smaller remnant of this type of abandoned quarrying can be seen about 100 feet above the old lime-kilns.

Cheltonians brought up with the Chimney as a prominent landmark in their lives may be surprised to learn that it is actually eroding away quite rapidly. Changes can be observed in it even over a limited period of a few years. It is eroding most rapidly on the

FIG. 58

DEVIL'S CHIMNEY
view from below

A quarryman's joke
made about 1780-90
All of the structure is in
THE LOWER FREESTONE

fragment of current bedded
rubbly limestone

two thick blocks of freestone

A major fissure
penetrating the Chimney
This will accelerate its
downfall within the next
25 to 50 years !

Pedestal basement
of closely jointed
freestone provides
strength to the
super structure

Scree slopes
colonised by
small Ash
trees

Pea Grit at
base of
Chimney

western side and the effect of frost is gradually disintegrating the rock along all the joints.

ONION WEATHERING

All round the quarry it will be noticed how the rock flakes off, like the skin of an onion. This process is known as 'exfoliation' and it is caused by the action of frost. Masons call it 'spalling', or 'onion weathering'.

The Oolitic limestone is a very porous rock and often the water comes to the surface again (by capillarity) and then freezes. Frost then splits off the egg-shaped ooliths into layers. Many layers are peeled off by severe frosts in winter.

Quarrymen used to say that Freestones like the Cheltenham Freestone should be cut into the required shapes when 'green', i.e. damp. They would explain that water exuded from the stone, and that when this evaporated a film of carbonate of lime would be left on the faces of the stone. They said it was wrong to scrape the stone when it was in place in a building in order to obtain a uniform appearance because this removed the protective film.

The whole process of onion weathering is accentuated in a town. The sulphur dioxide from oil fumes and coal fires combines with oxygen and water to form sulphuric acid which, in turn, attacks the calcium carbonate, setting free carbon dioxide and forming calcium sulphate. This peels off, leaving layers of soot behind, and is the reason why so many Regency buildings in Cheltenham look as weathered as any medieval building although actually built in the late eighteenth and the early nineteenth centuries.

It will also be noticed that in some parts of the quarry the rocks are black with what looks like atmospheric pollution but is actually a black lichen called *Verrucaria maura*. An orange-coloured lichen called *Xanthoria* occurs in sunnier parts of the quarry. Acids produced by both these lichens accentuate the erosion of the rock.

THE RAGSTONES

A close inspection of the main cliff-face will reveal that the rocks below are much better-jointed Freestone than those higher up. Lower Freestone is the name of the lower rock and the rock above is called the Upper Freestone and is about thirty feet thick. The Lower Freestone in this area attains a maximum thickness of 130 feet and forms the main mass of the escarpment of the Cheltenham area. Both divisions can be identified by their fossil content.

In between the Upper and Lower Freestone lies a band of rock which is not at all like the proper oolite. It is about ten feet thick and is called the Oolitic Marl—and can be best seen in the quarry face behind the Devil's Chimney. It is a white chalky rock which weathers into rubbly masses because it is not jointed strongly. The dominant fossil in this rock is a type of shell called *Terebratula fimbria*, or, to give it its more recently-acquired name, *Plectothyris fimbria*.

Above the Upper Freestone lie the Ragstones, closely-bedded limestones with the bedding planes only one or two feet apart. This makes it an excellent rock for the dry-stone walling which is such a pleasant feature of the Cotswold landscape. It is also very handy to get at—just dig a hole in a field!

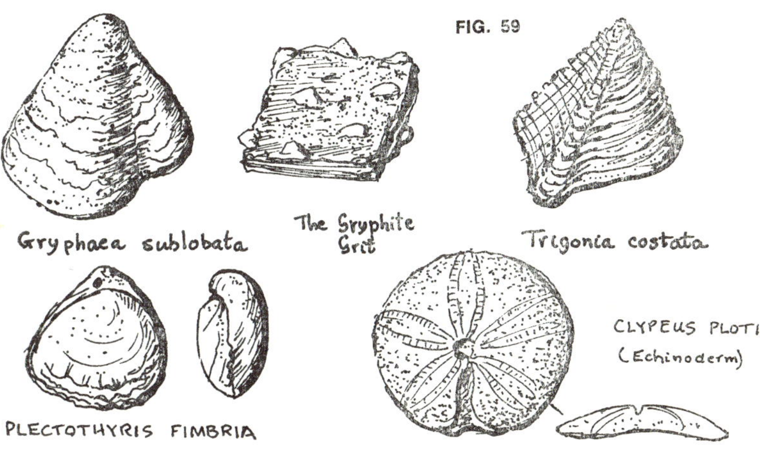

The stone wall extending between the Devil's Chimney and Salterley Grange Quarry has several holes alongside it. These resemble bomb-craters but are actually places where stone was dug out for wall-making during the Middle Ages.

The Ragstones are very shelly limestones to which the name 'grit' has been given—although they are not grits according to the geological meaning of the word. They are very fossiliferous and the different zones of grit have mainly been named after the principal type of fossil found in each zone. Two of these fossils are the bivalves *Gryphaea sublobata* and *Trigonia*. *Gryphaea sublobata* is a

G

large oyster type of shell which is the most common fossil in these rocks and very conspicuous. It weathers white and so the stone of this zone makes excellent ornamental walls and rock gardens.

Ragstones were very popular in Cheltenham in the middle and late nineteenth century as suitable stones for the Neo-Gothic buildings which were the architectural fashion of that time, and certainly the ragged nature of these stones does make such buildings look truly medieval—they are pseudo-Gothic but pleasantly and successfully so and an agreeable contrast to the Greek Revival and Georgian buildings, where blocks of Freestone have been used.

On the top of the cliff and behind the Iron Age earthwork camp are old quarries exposing the 'Grit' Ragstones to best advantage, with the best fossil finds also, *Trigonia* casts being very large and conspicuous.

SALTERLEY GRANGE QUARRY

Over the hill and not far away from the Devil's Chimney is Salterley Grange Quarry (see Map 9)—small and not so well-known as Leckhampton Quarry but nevertheless a geologist's paradise because the main features of the rocks show up so clearly and neatly. This is heartily recommended to all colour-photography

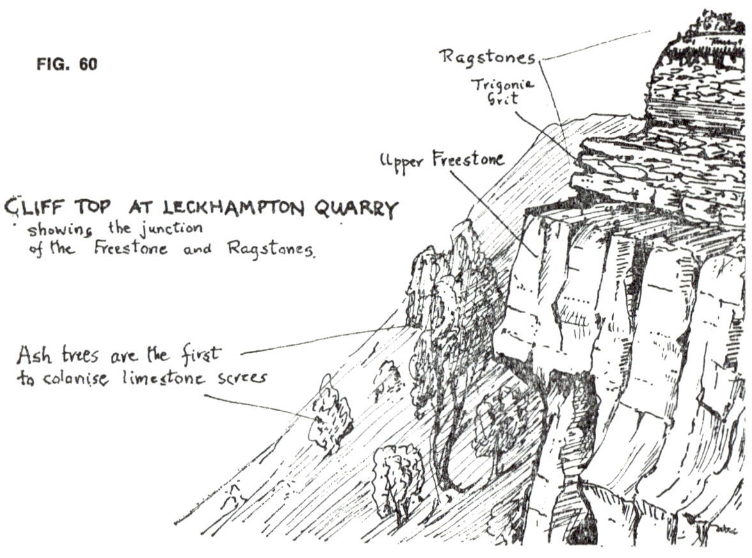

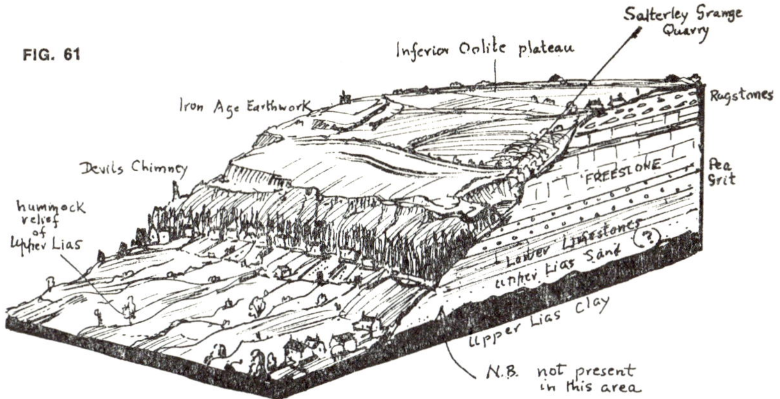

enthusiasts because a photograph of this quarry in afternoon sun-shine would be one of clean, cream-coloured rocks showing clearly-defined features of bedding planes, joints and fissures.

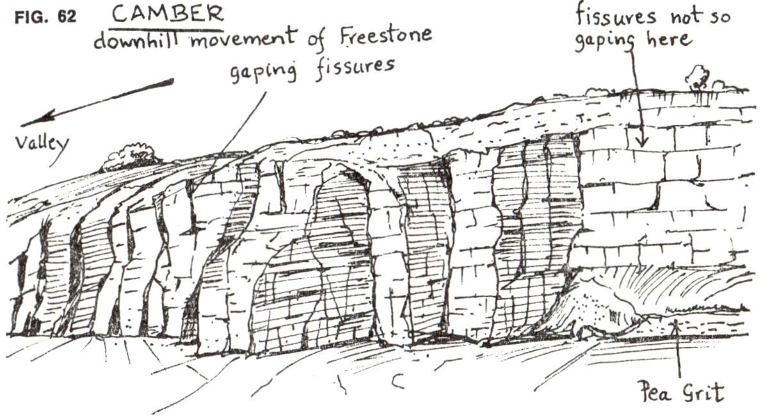

Salterley Grange Quarry — Lower Freestone.
On the left the rocks are gradually slipping downhill thus widening out the fissures. The limestones rest on a weak foundation of sands and clays.

The floor of this quarry is the Pea Grit and the cliff-face is of Lower Freestone.

Crystals of calcite can be collected in the joints of the Freestone, which are often covered with a deposit of calcium carbonate from percolating water, this being a typical cave deposit called Travertine, a formation which is yet another kind of rock which can be used for building when available in large masses.

In some parts of Italy, Travertine has been deposited in such vast quantities that this calcareous deposit from springs, which hardens on exposure, has developed into great masses of porous light yellow rock much used for building.

On the eastern side of the quarry is displayed a good example of a 'fault' (a fracture in the rocks with subsequent displacement of the rocks). This is made obvious by the fact that the Pea Grit appears on a shelf on one side and cannot be traced on the other. One interpretation of this is that it is just a reversed fault caused by compression; another is that the fault is due to camber, i.e. to rock masses slipping downhill.

CHARLTON KINGS COMMON

This cliff-top walk is one of the finest in the area for its commanding views across the Severn Vale and Cheltenham. There is a public footpath all along the cliff edge.

The first thing to notice is the large number of stone walls built

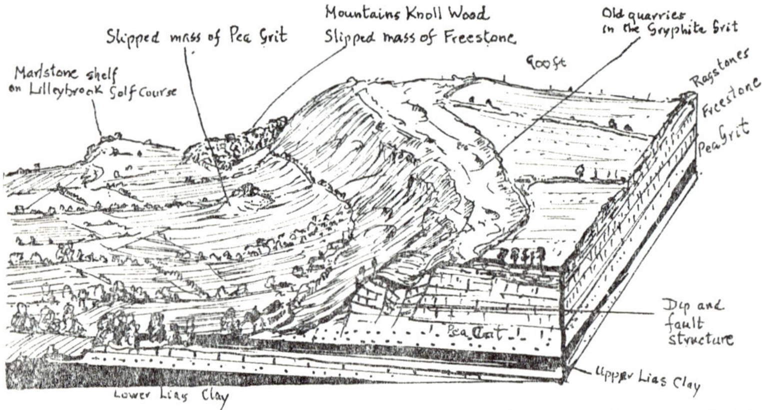

BLOCK DIAGRAM OF CHARLTON KINGS COMMON — looking east   FIG. 63

by the farmers. Obviously the Ragstones must be near at hand, and
they do, in fact, form the main capping rock of the 900-foot plateau
of the Cotswolds.

These shelly limestones are quite hard, harder than the Freestones,
which is one of the reasons why the Cotswolds form a high, sloping
plateau. Generally speaking, the harder the rock the higher the hill,
more resistant rocks standing out as higher ground.

If the Cotswolds were merely made of the softer Freestones,
erosion would have been much more rapid and a lower plateau
would have been the result. As it is, the Clypeus Grit—named after
the sea urchin echinoderm, *Clypeus ploti*—forms an extensive cover
to the high Cotswold plateau in this locality, whereas elsewhere
other Ragstones form a hard, resistant capping to the plateau.

At the top of Charlton Kings Common there are numerous old
quarries in the Gryphite Grit offering a good hunting-ground for the
fossil oyster *Gryphaea sublobata*.

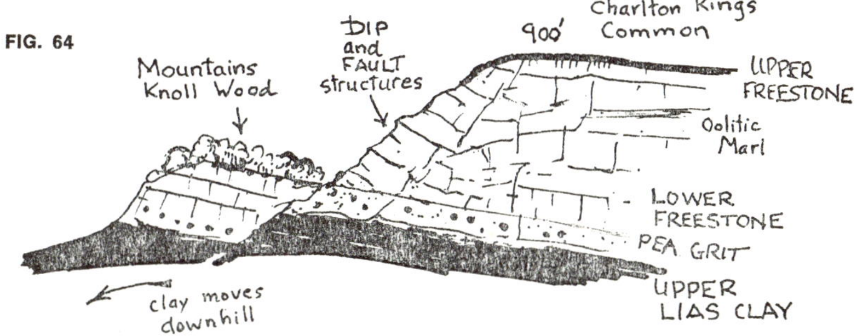

Mountains Knoll Wood at the eastern end of Charlton Kings
Common, is a slipped mass of Oolite. Notice the false
dips due to the rock masses resting on slippery clay.

Keen geologists will observe that on small promontories at each
end the rocks begin to slip and tumble. Although the limestones dip
south-eastwards about one degree they appear to be horizontal—yet
at each end of the common dips can be observed of as much as
sixty to seventy degrees. This slipping down the hill of masses
of limestone results in small faults and steep dips, processes
known as 'slumping' and 'cambering'. Both can be seen in an incipi-
ent stage along the path at the eastern end of the common, where

the beds are turned sharply upwards and Mountain Knolls Wood is a slipped-down mass forming a small outlier.

A small knoll near Sandy Lane and above Southfields Farm is a mass of Pea Grit, a few hundred square yards of displaced rock.

There are not very many springs in this area, but one good one supplies Black Hedge Farm and another feeds Southfield Brook and Southfield Farm (see geological sketch Map 10). This comparative dearth of springs is because the main mass of the rocks dips to the south-east, away from the main escarpment—and as the scarp is very close to the drainage of the Thames, most of the underground water travels down the dip to the Thames.

There is an Iron Age camp on the top of Leckhampton Hill, and there can be no doubt that the geology of the hill determined this choice of site.

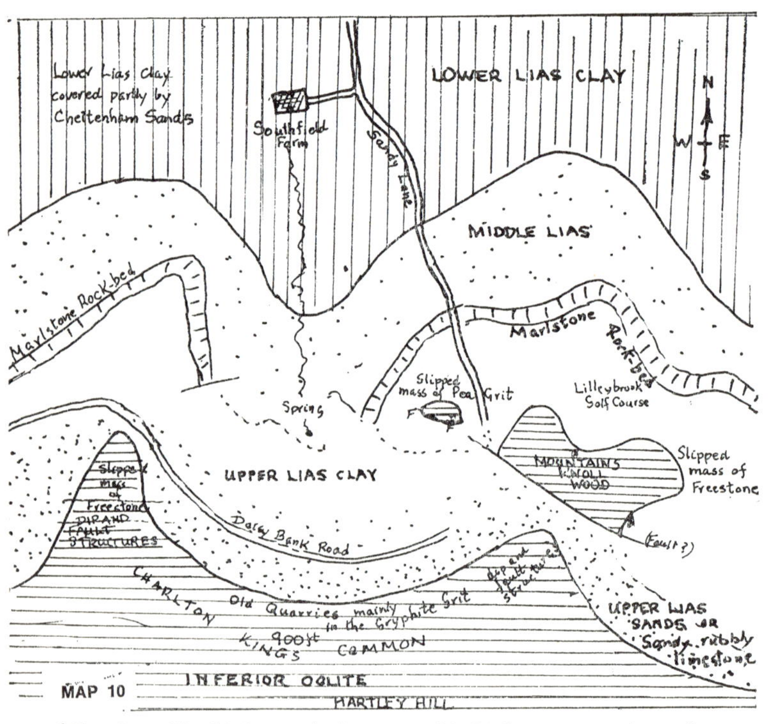

GEOLOGICAL SKETCH MAP OF CHARLTON KINGS COMMON
Scale 6" — 1 mile

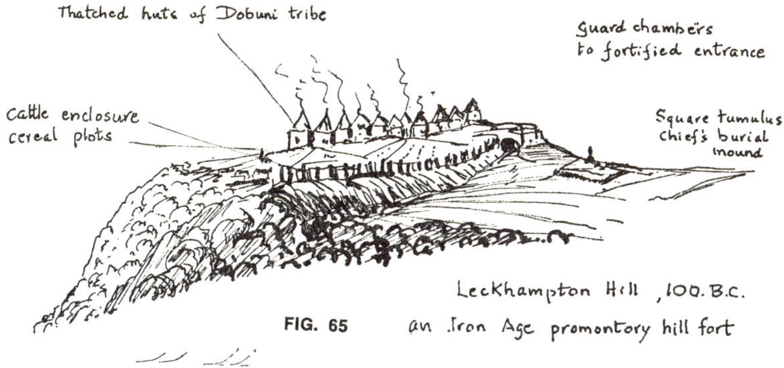

FIG. 65

Taking advantage of the promontory of the hilltop, the Dobuni tribes which occupied this camp had only one major defensive earthwork to build to make the place safe to live in and they used the Ragstones to form the core of the wall (which is now exposed in a few places).

This site is also favourable for signalling to other camps (Bredon, Oxenton, Dixton, Nottingham Hill, Cleeve, Battledown, Crickley, Birdlip, Painswick, Coopers Hill and Churchdown camps) and is in a direct line with most of the ancient trackways south.

The camp was excavated in 1925 but all that can now be seen is an earthwork rampart guarding an enclosure. The whole camp occupied an area of about eight acres and stood at a height of 965 feet. There was a fortified gateway and the excavations revealed two guard chambers. Outside the camp is a curious square tumulus.

These people knew how to cultivate the soil, domesticate animals and make pottery, a quantity of which was found on the floors of both guard chambers. They later saw the Roman soldiers from the garrison at Glevum (Gloucester), traded with the Romans and worked on their villa estates. But, before the Romans came, they lived in settlements somewhat like the sketch above, which is of Leckhampton Hill as it might have been in 100 BC.

CHAPTER 12

Cleeve Hill

The great mass of Cleeve Hill lies midway between Winchcombe and Cheltenham—and for bracing air, glorious views and a dry walk in the wettest weather it is the best place in the Cotswolds.

This is the highest part of the Cotswolds (the summit at the Ordnance Survey Trig. point is at 1,083 feet), and many people feel that this is also the area with the most spectacular scenery—secluded upland valleys, rather like Alpine valleys in summer, as well as breathtaking views from the escarpment.

FIG. 66   CLEEVE HILL — scenery like parts of the Pennines

Cleeve Hill is also the highest point of the Jurassic strata which dip down to the plain of Oxford, and here the Inferior Oolite limestone of the Middle Jurassic is at its thickest, thinning out southwards towards Bath, where the Greater Oolite is much thicker and overlies it.

Figure 67 is a block diagram of the main mass of Cleeve Hill, and Figure 68 shows the underlying structures in rather more detail. Notice that there appears to be a break in the sequence and disposi-

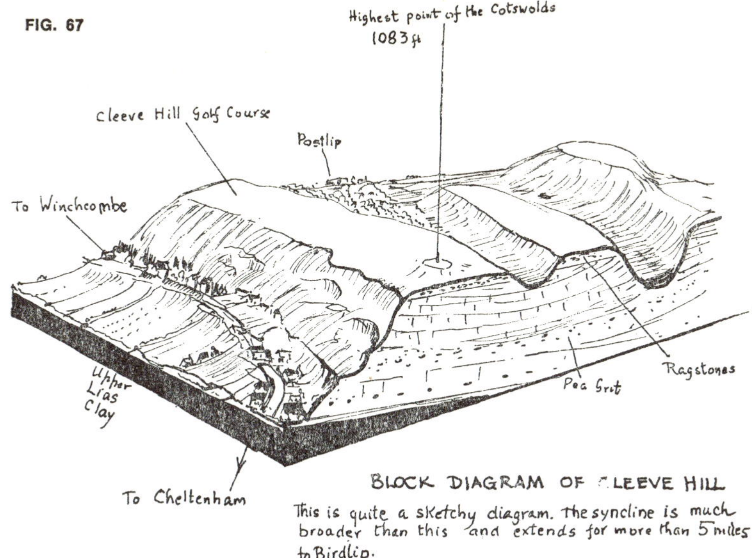

BLOCK DIAGRAM OF CLEEVE HILL

This is quite a sketchy diagram. The syncline is much broader than this and extends for more than 5 miles to Birdlip.

tion between the Upper Trigonia Grit and the other limestones below. This 'unconformity' tells a geologist that some of the limestones of the Lower Inferior Oolite were folded and slightly eroded after deposition and compaction, after which the Upper Trigonia Grit was laid down on top, *horizontally*.

The first part of the story dates back about 180 million years ago, when a basin, or 'syncline', extended from the Mendip axis right across to the valley of Moreton. From Middle Lias times right

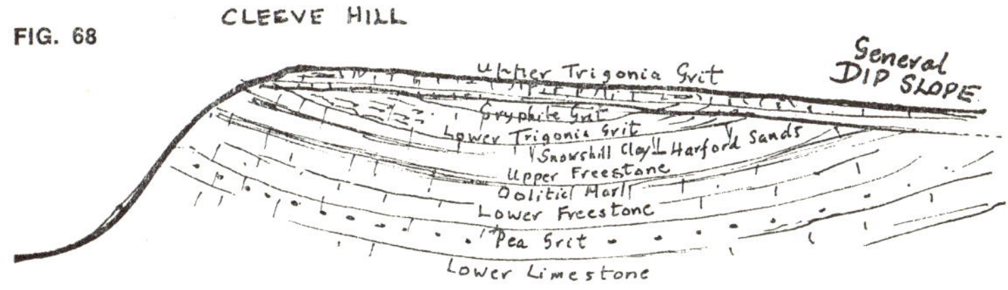

Simplified diagram explaining why the limestones of the Inferior Oolite reach their maximum thickness on Cleeve Hill

The strata dip south east at about 1° or a fall of 70 ft to 1 mile

up to when the Ragstones were laid down, a great mass of rocks was deposited in the basin, to a total thickness of some 800 feet. The centre of this syncline was in the Cleeve Hill area.

The coastline during this time was somewhere on the Welsh borderlands and in the Cleeve Hill area the sea was clear and warm, with corals, crinoids, bivalves and brachiopods flourishing in abundance. Thus, to look at the limestones in the numerous quarries all over Cleeve Hill is really to look at the debris of calcareous seas and, in some places, coral reefs. Sometimes the seas became sandy, at others they were muddy—and each condition brought with it a different assemblage of marine organisms.

Where deposition was continuous over a long period the bedding planes are further apart and the limestones are far more massive. It should be remembered that although the Ragstones cover the main hill-mass, it is the Notgrove Freestone of the Middle Inferior Oolite, fifteen to twenty-five feet thick, which covers a wide area in the North Cotswolds.

THE POSTLIP VALLEYS

The explanation of why the scenery is so dramatic covers a period much later than the actual deposition of the rocks on Cleeve Hill.

On a walk across the golf course to the Postlip valleys the landscape presents a wide expanse of downland with deep valleys leading down to Postlip. The writer believes that these were formed under periglacial conditions, either when the rainfall was much higher or when there was much surface water because the sub-soil was frozen.

Wherever a valley is deep enough to reach the present level of underground water a spring can be found, and at Postlip, above a small pond known as the Washpool, a most interesting spring occurs at the junction of the Upper Lias sands and the Upper Lias clay (Figure 69).

A vast amount of water is locked up in the Cotswolds because the porous oolite rock soaks it up like a sponge. The average rainfall today for this area is about thirty to thirty-five inches (the higher the ground the heavier the rainfall) and one inch of rain is equivalent to $101\frac{1}{4}$ tons of water per acre. Even leaving out the loss due to evaporation, much of this water reaches the Upper Lias sands through the numerous vertical joints and fissures in the oolite. Both

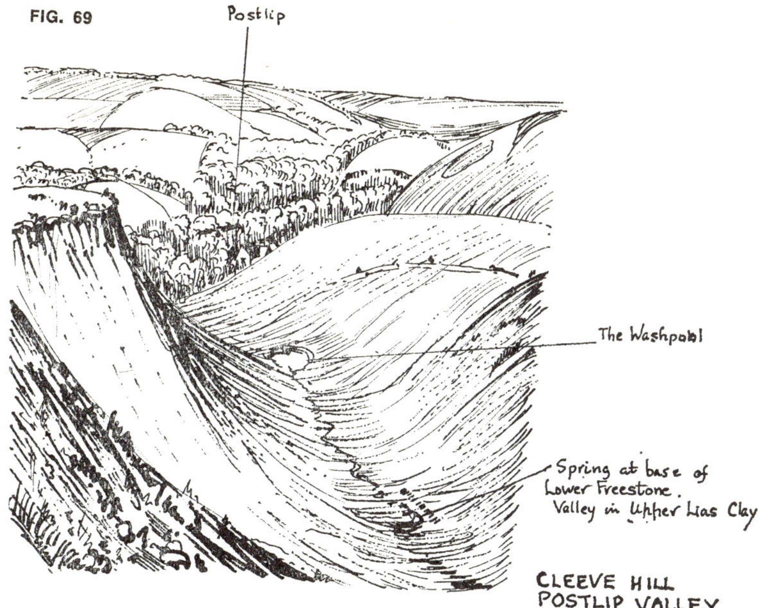

the sands and the limestones act as vast natural reservoirs—always releasing the water with constant flow. Very few of the springs here dried up in the great drought of 1921.

Owing to the structure of the Cleeve Hill mass, the more copious springs are located on the south-eastern side, and at Syreford, near Andoversford, there is one particularly powerful spring from which a pumping-station delivers water to several nearby villages.

Many farms in this part of the Cotswolds are located at or near the spring-line junction between the Upper Lias clay and Upper Lias sands but, of course, it is far better to have the actual farm buildings on dry land, i.e. on the sandy beds, with the spring below. Such farm settlements use a hydraulic ram to get the water up to the farm.

An interesting occupation for ramblers is finding the locations of springs on the Cotswolds and studying their relationship with the strata. The one-inch-to-the-mile geological map of each area should be used, and in the case of Cleeve Hill this is the Moreton-in-Marsh Geological Survey Sheet.

Most of the springs will be found to be petrifying in some degree

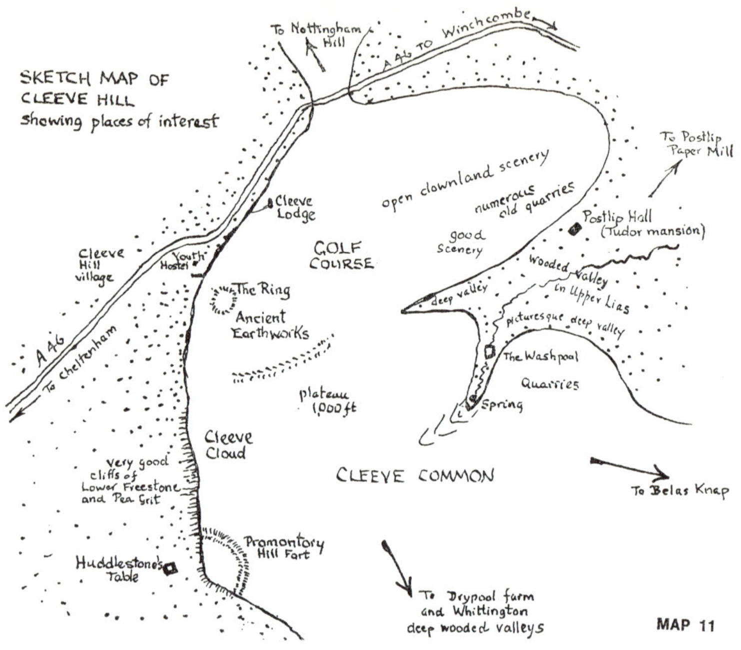

CLEEVE HILL AND NOTTINGHAM HILL

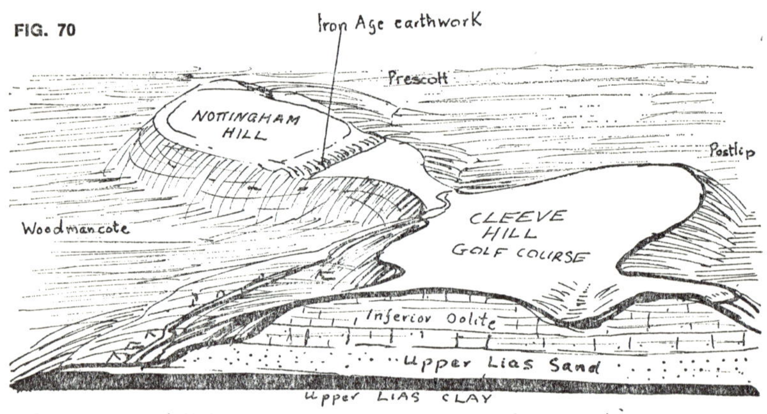

Promontory hill fort of NOTTINGHAM HILL (Iron Age)
The Dobuni tribes found that only the minimum defences were necessary if they used a Cotswold promontory.
Only one earthwork was needed, as at Leckhampton Camp

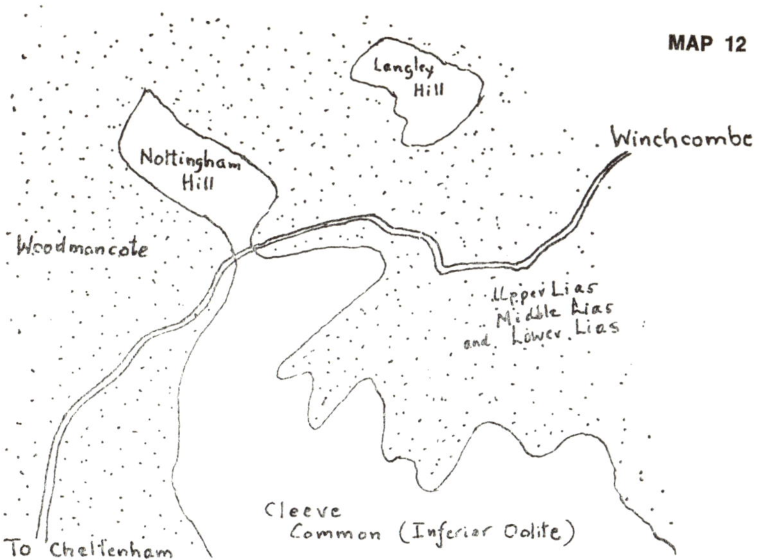

GEOLOGICAL SKETCH MAP OF THE CLEEVE HILL AREA

Langley Hill is an outlier and Nottingham Hill is rapidly becoming one. A small "isthmus of Inferior Oolite" joins it to the main plateau; it provides a useful "col" for the main road from Cheltenham to Winchcombe.

and frequently twigs and leaves covered with tufa can be discovered, or even petrified moss and grass. The water is hard but very clear and it was this clarity that determined the establishment of the small paper mill at Postlip. This makes a special kind of filter paper and has been a going concern since the eighteenth century.

This hill mass forms rather a conspicuous promontory when viewed from the Cleeve Hill golf course. Figure 70 shows how it has almost become an outlier, for only a thin neck of Inferior Oolite joins it with the Cotswold plateau.

Normal erosion has completely separated Langley Hill (Map 12) so this is technically an outlier, although joined on to the Cotswolds by the surrounding Middle Lias.

Again, on Nottingham Hill, Iron Age tribes made use of the strategic position of a promontory merely by making one earthwork

FIG. 71

CLEEVE CLOUD. exposure of Pea Grit and Lower Freestone best example of current bedding in the region. Other examples are at Barrow Wake

along the 'isthmus'. The result here was a defensive site covering over 100 acres, one of the largest in the county. Occasionally pieces of flint scrapers have been found (winter being the best time for such discoveries as there is then less vegetation covering the ground) but it should be remembered that flints originally come from chalk country. The rambler who finds them on the Cotswolds does so because Man brought them there.

CLEEVE CLOUD

South of Cleeve Hill Youth Hostel, the rambler can follow the line of old quarries along the foot of the cliffs towards Huddlestone Table, a rectangular mass of dressed stone just below the Iron Age earthwork on Cleeve Common. Parts of the Lower Freestone here are quite massive and provide relatively safe faces on which to practise rock-climbing with ropes.

The Pea Grit formation here is just as massive as at Crickley Hill, although it is not pisolithic throughout.

Figure 71 shows parts of the Pea Grit overlaid by a fine example of current bedding in the Lower Freestone.

The whole area of both Cleeve Cloud and Cleeve Common is riddled with old quarries, medieval trackways and pre-Roman earthworks.

Chapter 13

Barrow Wake, Crickley Hill and Birdlip

Barrow Wake, just past Birdlip on the Cheltenham road, is undoubtedly one of the most spectacular viewpoints of the Cotswolds. Even the snack-bars and the litter which bear mournful witness to this cannot lessen the glory of its panoramic views.

There are, inevitably, geological reasons for this spectacular scenery, for this is the edge of the Cotswold escarpment some 900 feet above sea level, and the view stretches right across the Vale of the Severn to Wales and the Malvern Hills. Diversity is added to the scene by the outliers of Robin's Wood Hill and Churchdown Hill, which stand like sentinels guarding the approaches to Cheltenham and Gloucester.

FIG. 72

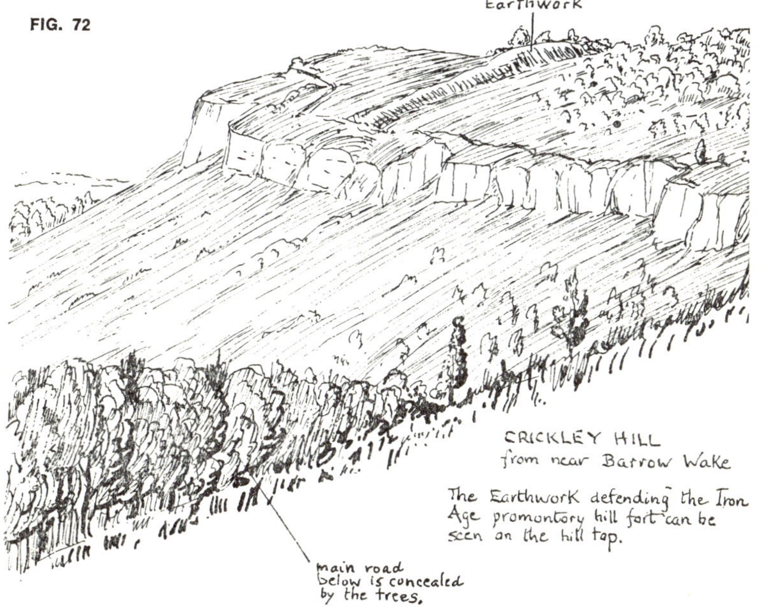

The wide vistas are helped by the absence of woodlands in the immediate vicinity, the result of extensive quarrying, and, towards the Air Balloon Inn, the view extends to the promontory of Crickley Hill with its castle-like cliffs, on one of which an Iron Age earthwork stands out clearly. 'Earthwork', though, is rather a misnomer for this type of construction because, although the surface is of earth, it has actually been built by piling up rocks, which can be seen merely by digging a few inches below the surface. Advantage of the local rock was also taken in the actual siting of the fort and by using the escarpment itself as a barrier the rampart did not need to extend right round the fort.

CRICKLEY HILL

The cliffs on the west side of the Crickley Hill promontory form the finest exposure of the Pea Grit formation in Britain—another geologists' mecca!

The extraordinary deep valley between Crickley Hill and Barrow Wake is partly due to a fault, which can be verified 'in the field' by

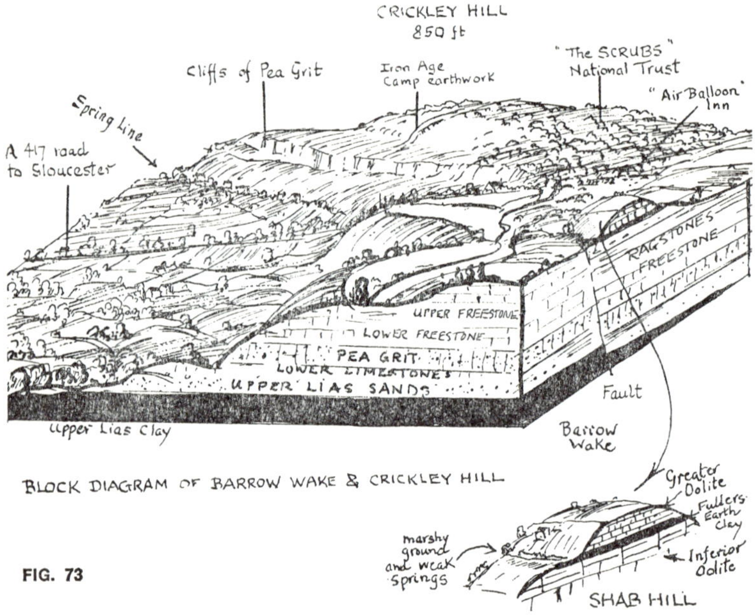

FIG. 73

CRICKLEY HILL   Massive Pea Grit formation
The widely spaced bedding planes indicate continuous
and rapid sedimentation.

examining the small quarries on either side of the road to Barrow Wake. Notice also the curious fact that, near the Air Balloon, the Upper Inferior Oolite outcrops with the exposures of the various Ragstones, yet higher up the hill the Upper Freestone outcrops.

A small spring on the high ground near Shab Hill (see the block diagram at Figure 73) indicates the presence of a band of clay, known as the Fuller's Earth, which forms high spring levels in many places on the Cotswolds. Above this occur the limestones of the Great Oolite which will be more fully discussed in the next chapter.

On one part of Crickley Hill, immediately across the intervening

H

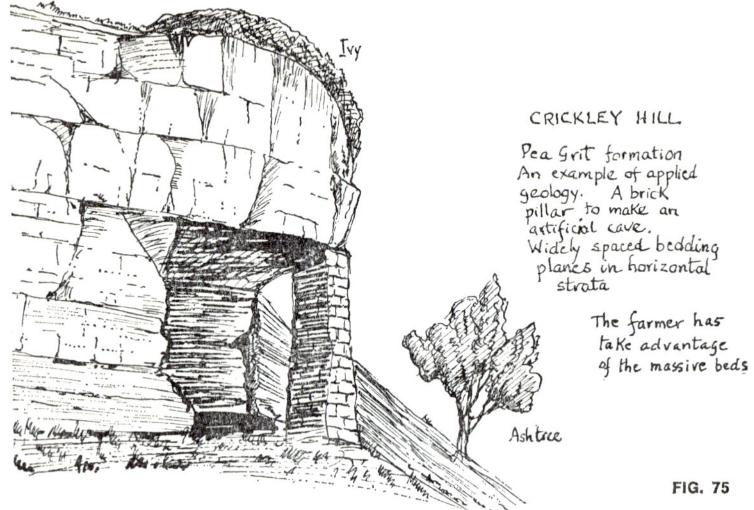

FIG. 75

valley, there is a cliff walk along magnificent masses of the Pea Grit (see Figure 74), the formation here reaching its maximum thickness of forty feet.

Figure 75 shows a humorous example of the exploitation by man of geological features in this formation. The horizontal bedding planes of the massive rocks form the ceiling of an artificial cave, the ceiling of which is supported at the corner by a man-made pillar.

After crossing the well-preserved earthwork ramparts, the geologist will notice that the Pea Grit cliffs at this particular point become very fossiliferous, the rocks teeming with fragments of sea urchins, polyzoa and various brachiopods.

The final point of interest at the extreme end of the hill is the Devil's Table, a naturally-formed pedestal caused by erosion in the horizontal Pea Grit formation, and, conceivably, used for religious ceremonies by the inhabitants of the Iron Age camp (Figure 76).

The quarry on the north side of the promontory is merely extracting stone to be crushed into road metal but is worth a visit to see the varying dips caused by cambering and the display of master joints which are major joints traversing the oolite across long distances.

Crickley Hill is a good viewpoint from which to see the bay-like recesses which have been eroded into the scarp edge. Some are large (e.g. Winchcombe) and some small like Witcombe. They are usually floored with Lower Lias clay which provides good pasture, the green of which stands out in contrast to the wooded slopes of the oolite.

BIRDLIP

Little mention has been made so far of the Ragstone divisions of the Upper Inferior Oolite, and to recognise these in the Cotswolds calls for a good deal of local field-work. Easiest to recognise are the Clypeus Grit and the Trigonia Grit, each identifiable by their respective fossils.

At Birdlip, on the edge of the scarp and only a few hundred yards downhill from The Royal George Inn, the Upper Trigonia Grit can be seen resting on Upper Freestone. This means that many of the other Ragstones are missing from this crest of an underlying structure known as the Birdlip anticline, which stood above the sea when the Ragstones were being deposited in the shallow water around it. It is likely that the Birdlip anticline was uplifted at the end of Middle Inferior Oolite times and that the Ragstones which had been deposited everywhere were then eroded from it.

FIG. 76

THE BIRDLIP ANTICLINE

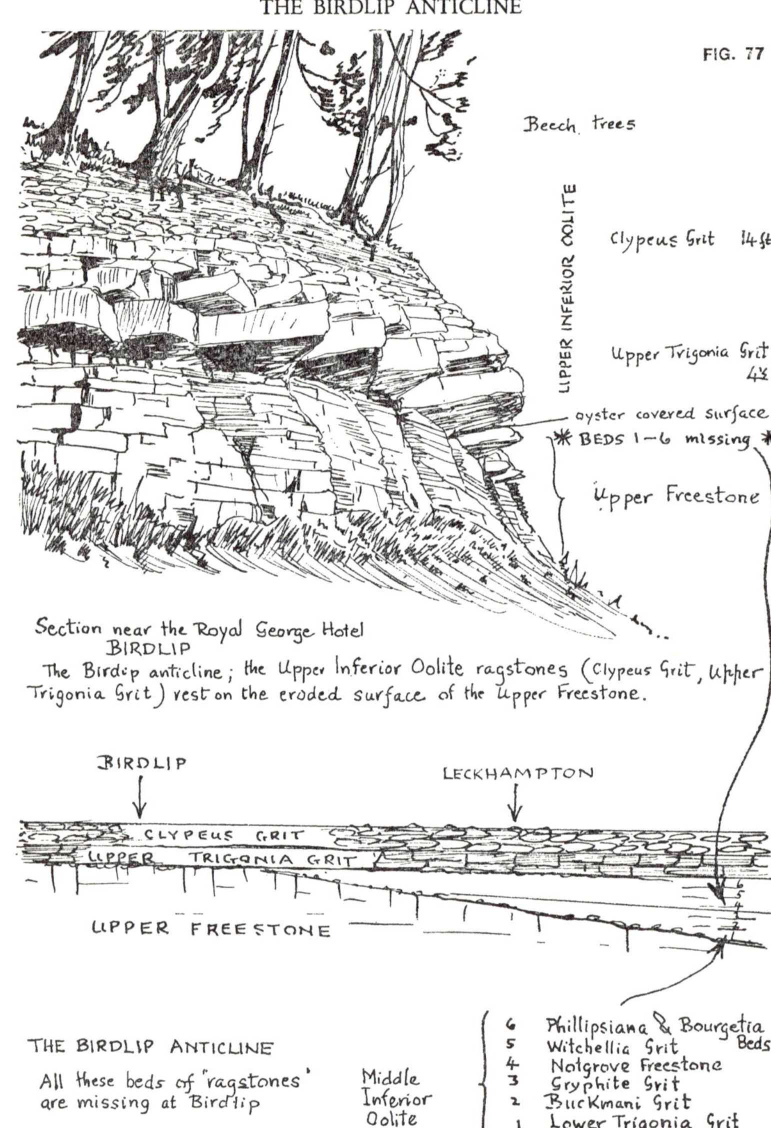

THE VIEW FROM THE PEAK

At this point on the scarp there is an excellent vantage point known as The Peak, from which there are views across the Vale of Witcombe to Gloucester and the outliers of Robin's Wood Hill and Churchdown Hill. Beyond Gloucester is May Hill, where the rocks are much older than the Jurassic of the Cotswolds, going as far back as the Silurian period.

The lower limestones form the base of the Inferior Oolite outcrop on the side of the Peak. The hill itself is capped by Pea Grit but, higher up, occur Freestones capped by Upper Trigonia Grit and Clypeus Grit.

Figure 77 shows a sketch of this fine cliff which, topped by beech trees, can be seen on the escarpment only a few hundred yards from The Royal George Inn at Birdlip. Notice that all the beds numbered from one to six are missing from this cliff, yet they appear on Cleeve Hill. Evidently the beds were here originally but were eroded away after they had been folded into an anticline at Birdlip and a cyncline at Cleeve Hill.

All these structures have been most painstakingly analysed by the nineteenth-century geologist, Buckman, and during this century by L. Richardson and Dr D. V. Ager, but amateur geologists paying brief visits to these sites should not be disappointed if they cannot always find what is reported in the textbooks. The sites in this area are most complicated geologically and to understand them fully it is necessary to have a good knowledge of local quarries and railway-cuttings, as well as a very detailed knowledge of local fossils.

CHAPTER 14

The Combes

A 'combe', or 'coombe', is the general name for a short valley running into a hill. Along the Cotswold escarpment there are many such short valleys running into the scarp edge, often wide and shaped rather like part of an amphitheatre. They form bay-like recesses in the scarp edge.

The promontories which make up the sides of the combes often have an alignment lying WNW to ESE, and it is apparent that these promontories must have some relationship to the presence of faults in the Cotswolds.

A study of the one-inch-to-the-mile geological map reveals numerous faults in the Cotswolds and it is possible that many extend right into the Severn Vale. All these faults are divided into two sets—one trending to the west/north-west and the other mainly north/south.

The main group of faults in the Birdlip/Witcombe area and their

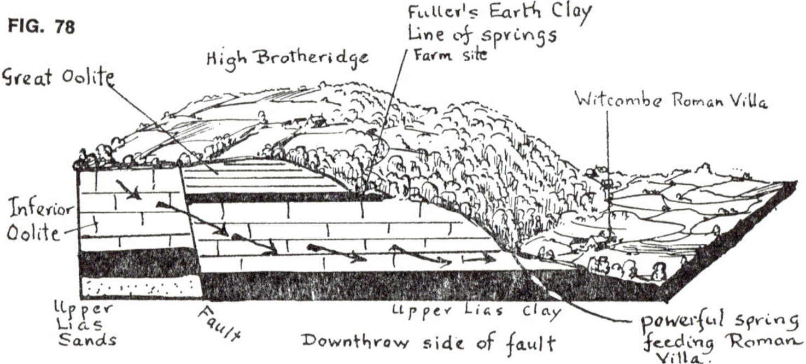

Arrows show escape of underground water to make for more powerful springs on one side of Witcombe

BLOCK DIAGRAM OF HIGH BROTHERIDGE

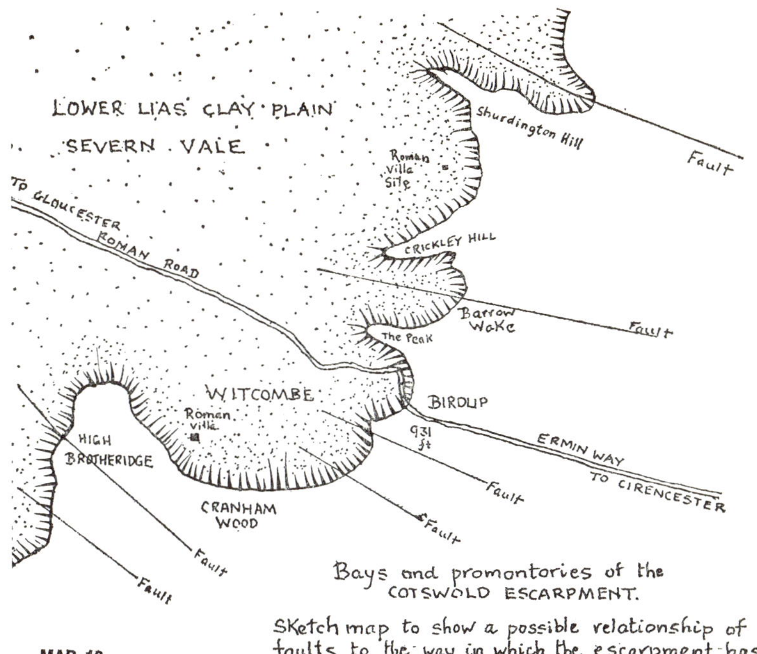

Bays and promontories of the
COTSWOLD ESCARPMENT.

Sketch map to show a possible relationship of
faults to the way in which the escarpment has
been eroded.

MAP 13

possible relationship to the way in which the escarpment has been eroded is shown in Map 13.

A fault, as we have seen, is a fracture in the rocks along which some movement has taken place, and if a fault should occur in the direction of dip across the escarpment it will throw the scarp forward. In the case of some small projections of the scarp, it may be that the topographical feature has been formed by slipped masses of oolite rather than normal faulting.

It is a paradox that downfaulting tends to stabilise the promontory, as it brings the harder rocks down to a lower level. This effect can be seen in the downfaulted blocks of Bredon Hill and Cleeve Hill, the outliers covered in previous chapters.

Another good example of a downfaulted mass which has stabilised the promontory can be seen at High Brotheridge (Map 14). This reveals a typical WNW fault bringing the limestones of the Great Oolite on to the high ground. See also Figure 78, which is a block diagram of High Brotheridge.

Between the Great Oolite and the Inferior Oolite is the Fuller's Earth clay, which forms a spring line on the hill at a high level, and these springs may well have influenced the Iron Age tribes in their selection of the site for a promontory hill fort, one of the largest in Gloucestershire.

This rather stump-shaped promontory forms a tabular hill about 900 feet high and occupies a commanding position jutting out into the Severn Vale. It is now covered with a large beech forest which, with Buckholt and Cranham Woods, are among the few remnant forests of the Cotswolds.

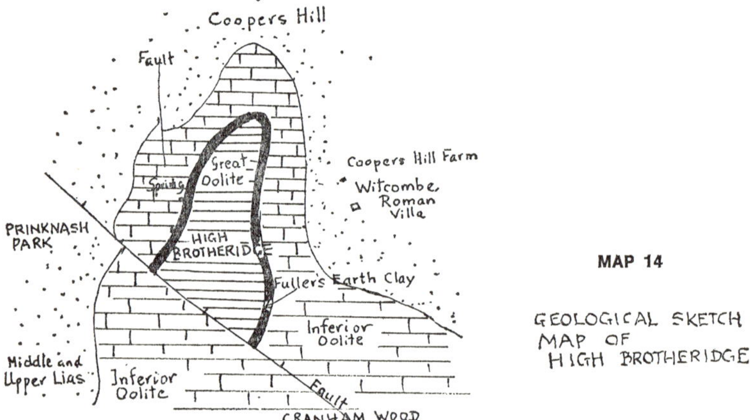

Within the Vale of Witcombe the geology varies with the height —the Lower Lias clay at 300 feet, the Middle Lias at 450 feet, and the Upper Lias up to 650 feet, where the limestones begin to outcrop (see Figure 79).

The semi-amphitheatre form of the combe is common to many along the scarp and it has been suggested that they might have been formed as a nivation hollow during the glacial period, each combe being a place where a snowfield collected.

The axis of the Birdlip anticline passes across the vale and this may have proved to be a line of weakness resulting in erosion, followed by erosion inwards alongside the promontories.

All round the combe it is often difficult to decide which are the precise boundaries of Lower, Middle and Upper Lias because the downwash material of oolitic debris and gravels has concealed the

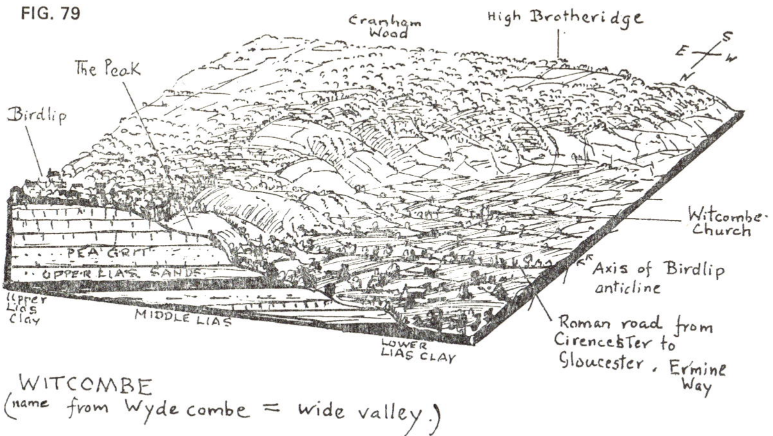

WITCOMBE
(name from Wyde combe = wide valley.)

outcrops. This tendency is accelerated by the presence of the Upper Lias clay which forms a lubricated surface on which rocks and detritus tend to slip downhill.

During the glacial periods (which often lasted tens of thousands of years) the sub-soil remained frozen to great depths and the frost-fractured rocks slid downhill when the upper layers thawed out. Inter-glacial periods lasted for many thousands of years when the climate was perhaps warmer and wetter than now.

This downhill mass movement is called 'solifluxion' and its associated unsorted gravels and sands are known as 'taele' gravels. When such deposits cover the slopes the landscape has been formed under periglacial conditions (see Figure 80).

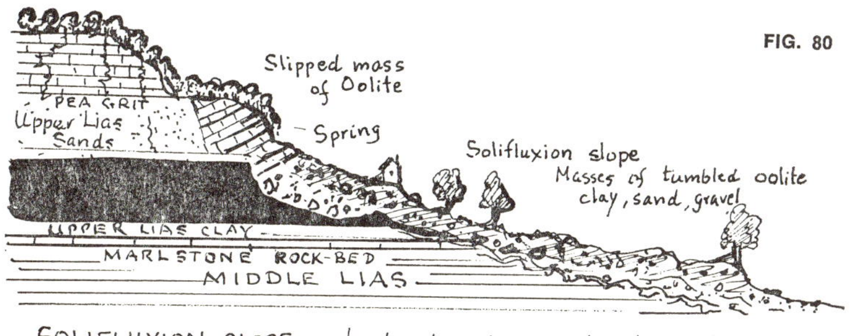

SOLIFLUXION SLOPE. — developed under periglacial conditions

There are many spurs on the combe slopes which were probably eroded under post-glacial and inter-glacial periods of much heavier rainfall than at present. Some of the spurs are due to minor faults or merely a slipping of the oolite. The combes, in fact, demonstrate the wide extent of slip in the Cotswolds.

The general orientation of the combes (opening out to the north-west) has had a marked effect on settlements. Villages tend to be mid-combe and the heavily-wooded slopes facing north are thinly settled. There is a resemblance here to the shadow and sunlight sides of Swiss valleys.

WINCHCOMBE—A LARGE COMBE

Sketch Map 15 shows how the vale at Winchcombe could possibly be due to erosion along a major line of weakness, the axis of a small anticline. It is very difficult to prove this suggestion but the combe has a north/south orientation and is not related to the system of parallels faulting to the south.

The River Isbourne flows northwards to join the Avon at Evesham. In the early Pleistocene period the Avon also flowed northwards to the Trent and was reversed to the Severn when a marginal glacial lake occupied the East Midlands. Thus the Isbourne represents a fossil drainage.

Much of the information about Witcombe applies to this area also but the landscape does show evidence of more powerful and extensive erosion here. This greater erosion was, perhaps, due to the vale opening out directly northwards towards the Ice Front when the ice-sheet lay across the mid-Avon area.

The Roman villa of Spoonley is related to a good spring issuing from the top of the Upper Lias clay but the Roman villa of Wadfield appears to have no nearby spring today, although in Roman times water was emerging at the surface. There is plenty of evidence that the level of underground water has been falling in the Cotswolds. Nevertheless there are still many springs around the combe, all of which feed into the hub of the wheel between Winchcombe and Sudeley Manor, so making the Isbourne quite a powerful little river.

There is a way out of the combe southwards by following up the Beesmoor valley to Charlton Abbots, where the Brook rises from a spring above the Upper Lias clay at nearly 600 feet. Only some 300 yards further on lies the source of the southward-flowing Coln in quite a shallow valley leading down to Brockhampton, Seven-

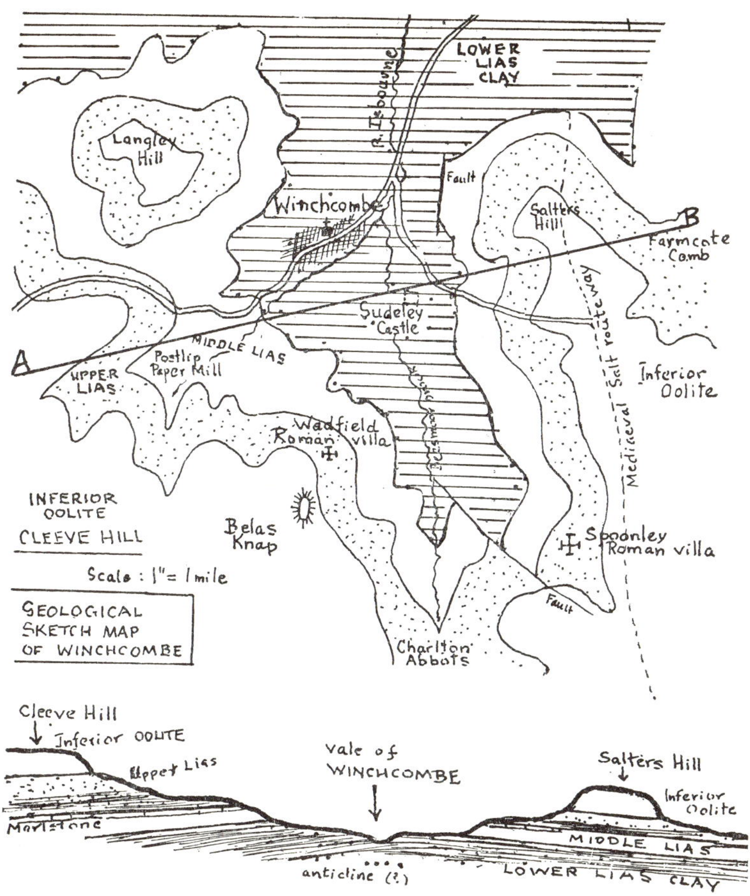

hampton, Syreford and onwards to the Thames. In contrast, the Beesmoor valley is deep, almost ravine-like, and the question of river capture here is much debated by geomorphologists.

A glance at the geological sketch (Map 15) will show that this Beesmoor (Isbourne/Avon drainage)—Coln valley system isolates the mass of Inferior Oolite of Cleeve Hill, making it one great outlier.

CHAPTER 15

Springs and Villages in the Great Oolite Regions

A large area of the mid-Cotswolds is covered with the oolitic limestone called the Great Oolite, which rarely exceeds a thickness of 100 feet. The Inferior Oolite, lying below but separated from it by a bed of Fuller's Earth clay, is also thinner here than in the North Cotswolds.

Figure 81 shows the usual order of succession in simplified form, the younger beds being on top. To understand the scenery developed on these rocks it is again necessary first to understand the nature of the rocks themselves.

THE GREAT OOLITE SERIES

The best sections are to be seen in the old railway-cuttings at Hampen between Andoversford and Notgrove. Andoversford to Bourton-on-the-Water, and between Withington and Chedworth, are particularly good stretches and the Gloucestershire Trust for Nature Conservation is hoping to take them over.

The Great Oolite is typically a hard, white, shelly limestone often showing rather strongly-developed current-bedding. This stone, which was used by the Romans as well as throughout the Middle Ages, can be cut with a saw when newly 'dug' but soon hardens on exposure to air.

One stratum in the series (again see Figure 81) is known as the Taynton Stone, and occurs in the Windrush, Barrington and Taynton areas. Locally, it was used for drinking-troughs and cottages but it has also been used for parts of important national buildings, including St Paul's Cathedral, though this is mainly built of Portland stone from Dorset.

At the base of the Great Oolite there occurs a series of sandy limestones which are quite fissile, i.e. easily split along thin planes. This is Stonesfield Slate and is discussed in the next chapter.

THE FULLER'S EARTH CLAY SERIES

This is mainly a clay formation which occurs all over the Mid- and South Cotswolds. In the north, it tends to disappear and is replaced by the Chipping Norton limestone forming the plateau levels between Temple Guiting and Upper Swell.

The clay is about 100 to 150 feet thick near Bath and some of it has the peculiar property of crumbling in water. A non-plastic clay, due to the presence of the montmorillonite type of clay mineral, it was widely used by fullers in the woollen industry for extracting grease from wool.

But in the Cotswolds, this is not a fulling clay. It is merely a

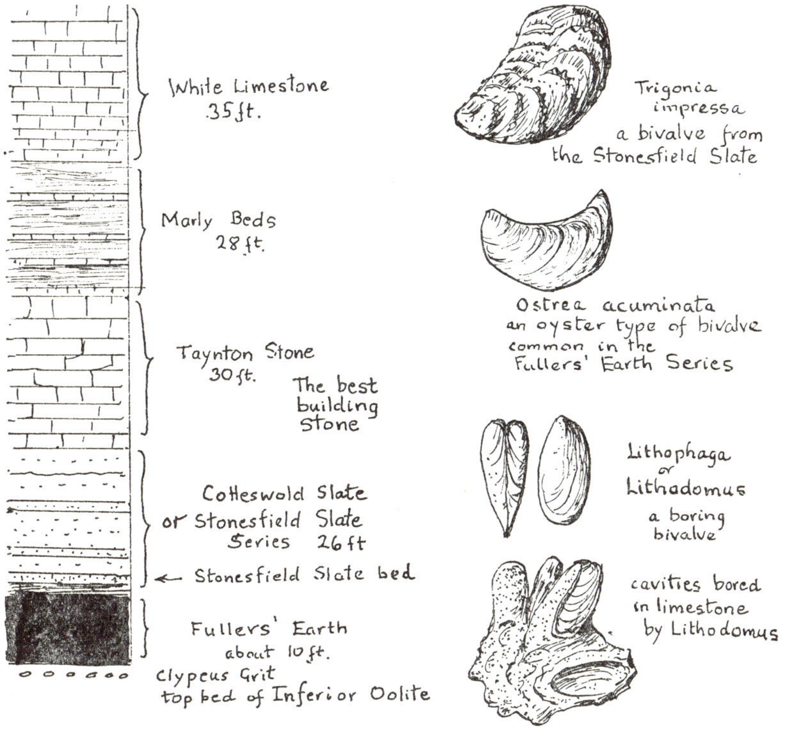

FIG. 81 The GREAT OOLITE SERIES as seen at the Hampen railway cutting between Notgrove and Andoversford.

band of clay varying in depth from fifty to seventy feet which acts as an impervious layer beneath the Great Oolite so that springs are thrown out at the base of the limestone.

Many villages and farm settlements were founded on the powerful springs released at the junction of the Fuller's Earth clay and the Great Oolite, and even up to the Second World War villages were still dependent on the village spring for their water. Unfortunately, however, the Great Oolite is often only thirty to fifty feet thick, which means that spring water can be polluted by garden refuse and sewage because most of it merely runs through the joints and crevices in the rocks and is not effectively filtered by the rock itself.

One village, however, Compton Abdale, still has a particularly copious spring, proudly marked by the villagers with a gaping crocodile mouth at the precise spot where the water gushes out.

THE 'LOST' VILLAGES

It is not generally known that there are quite a number of 'lost' villages in the Cotswolds, whose sites have only comparatively recently been discovered by means of aerial photography. The outlines of buildings which have long since disappeared can be seen with uncanny clarity from the air, and the colour of the vegetation growing over the old foundations will appear in the photograph as a different shade from that of the surrounding vegetation, even though there is no visible difference to an observer on the ground.

FIG. 83

It is generally believed that many of these 'lost' villages were deserted in the Middle Ages when the water supply failed, and the inhabitants migrated down to the kindlier valleys. Another cause was the enclosure of land by the owners, which began as a first phase in the Middle Ages. Interesting examples of these 'lost' villages are 'Manless Town' near Brimpsfield, Upper Blockley and Upper Coberley.

The correlation of geology with agriculture is well illustrated in the upland valleys in the Great Oolite. The high plateau, standing at over 800 feet, is good arable land planted with the most suitable cereal (barley), and the freshly-ploughed lands reveal everywhere a 'clitter' of limestone fragments. Down the hill but high up along the valley slopes, the rich green pastures developed on the Fuller's Earth clay can easily be detected, plus the sites of the numerous springs thrown out on the hillside. And, amateur geologists please note, a green band of more luscious grass round a hill shows the actual outcrop of the Fuller's Earth clay.

Below the base of the Fuller's Earth clay occurs the uppermost division of the Inferior Oolite, the Clypeus Grit. The typical fossil of this stratum is the sea urchin, *Clypeus ploti*, a great favourite with country children. Many specimens weigh just about a pound and it was an old custom of the market women at Stow-on-the-Wold to use them as butter weights, the stones being called 'pound stones'. A very good section of the Clypeus Grit can be seen in the old railway cutting west of Notgrove Station.

DRY VALLEYS

There are many valleys in the Cotswolds which are without rivers or merely have a tiny trickle of water, and yet much further down the valley there is a gushing stream issuing from a spring. A division in the valley is often found—a step dividing the valley into a higher upper one and a lower one.

There are several theories as to the origin of this phenomenon and a very plausible explanation is that, at the close of the Ice Age, the

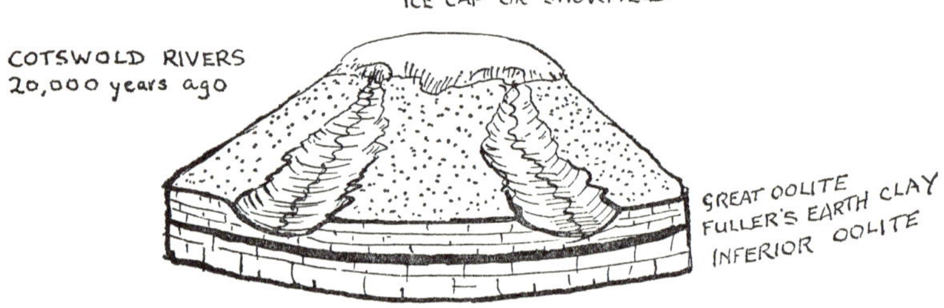

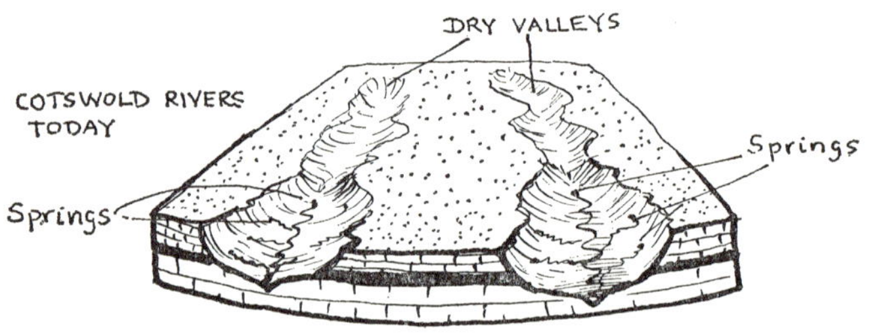

A THEORY ON THE ORIGIN OF DRY VALLEYS

FIG. 84

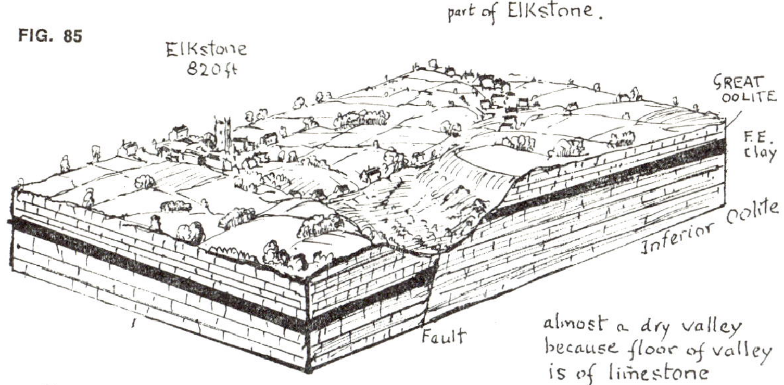

ELKSTONE – block diagram.
geology has exercised some control in making the village
into two settlements.

climate became very wet and the increased rainfall produced a
much higher water-table which has been falling ever since—hence
the 'lost' villages. Figure 84 shows how the River Coln illustrates
this theory by its large main 'fossil' meanders, along which the tiny
present-day meanders flow. The River Churn and other Thames
tributaries show similar features.

COTSWOLD VILLAGES

All the villages in the Cotswolds owe their origin to the geological
principles described in this chapter. Their siting and their construction have been directly controlled by the nature and disposition of
the rocks beneath them. No doubt that is why they are so beautiful
—they are completely in harmony with the landscape from which
they have sprung.

Some of them take a little finding—which is probably why they
have remained beautiful! The best approach is the A417 Roman
road known as the Ermine Way, which runs dead straight across
the somewhat monotonous plateau formed of the Great Oolite, with
numerous quarries here and there just off the road. This plateau was
nature's gift to the Romans for the site of a first-class military road,
the military road from Corinium (Cirencester) to Glevum
(Gloucester).

Occasionally the straight, ruler-like road dips into a hollow where
the lush green grass reveals the presence of the Fuller's Earth clay.

J

ELKSTONE. The Saxons built their villages above the spring line
The Romans preferred their villas below the springs
for the water to run through the baths

FIG. 86

These hollows mark the heads of valleys which strike across the
road and the traveller who follows a typical one three miles east of
Birdlip will turn off to the villages of Syde, Caudle Green and
Brimpsfield.

These are all villages at high levels (between 700 and 800 feet)
and yet the deep valleys are without villages. The dramatic depths
of the valleys and the mellow Cotswold stone walls of the cottages
perched high up on the plateau edge give the whole scene almost a
Mediterranean aspect in bright sunshine.

Elkstone is another pretty village with a beautiful Norman church
sited on the limestone but from which one looks down into a clay
hollow, the head of a valley draining down to the Churn and the
Thames. There are excellent springs below the village, and when the

FIG. 87

NOTGROVE
high village, 700 ft on the
Fuller's Earth Clay spring line

Fuller's Earth Clay
outcrop revealed by luxuriant grass
and cow pasture land

air temperature was 18° F in January 1963, the spring below the church steamed in the very cold air and watercress could be picked at the spring-head.

Further afield, there are the villages of Hawling, Notgrove and Turkdean, all high up but again with the unusual feature for such altitudes of cows grazing on rich meadows developed on clay. Above the villages, on the high plateau, there are large fields of cereals interspersed with beech copses used for pheasant breeding, though their main purpose is to act as wind-breaks.

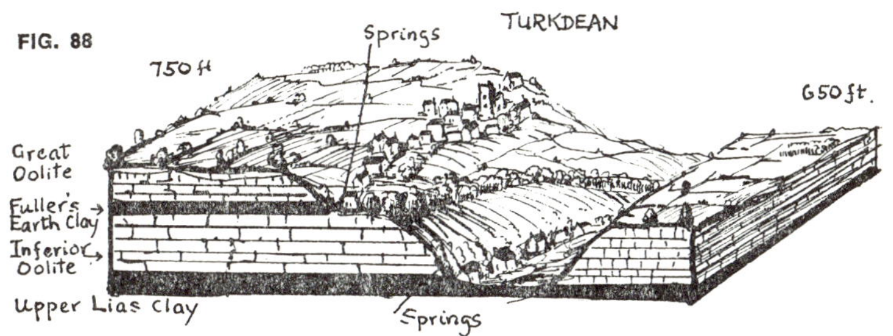

TURKDEAN.— high level village — at 650 ft. — based on high spring level

TURKDEAN — two settlements, valley and hilltop.

Turkdean (see Figures 88 and 89) is a very good example of two spring levels. The block diagram in Figure 88 shows how the valley has been eroded deep enough for the floor to reach the Upper Lias clay, thereby creating a second and lower spring line.

There are places in the Cotswolds—Elkstone is one example—

where the geological control is so marked that it has led to the splitting of the village into two settlements. One can trace how the peasants of the Middle Ages tried to preserve the economic independence of a village by developing areas with a variety of uses according to the nature of the geological outcrops. Whether a village is a nucleated village, a dual village or a linear pattern, it can usually be seen to coincide with the siting of the geological outcrops.

Chedworth is a good example of a linear village, with the more powerful springs occurring on one side of the valley—and an excellent place at which to study the relationship of clay and limestone and the way in which the village has grown along the lines of the strata.

It is also interesting to note that, since the railway line was closed, the Chedworth tunnel has had to be blocked up. It was driven through the rock almost at the junction of the Fuller's Earth clay and the Great Oolite and when railway maintenance workers ceased to check its condition, the seepage of water made it dangerous and it had to be sealed off.

Chedworth's famous Roman villa, extensive and obviously once very prosperous, is sited at the spring-line junction—a good water supply which is still working.

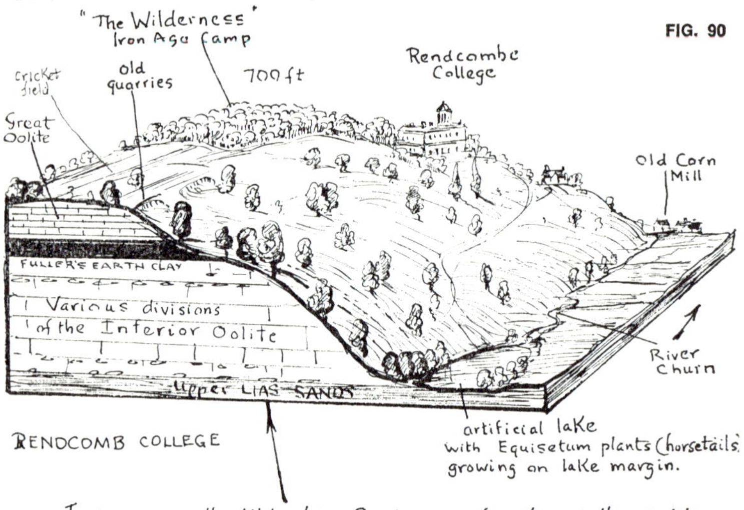

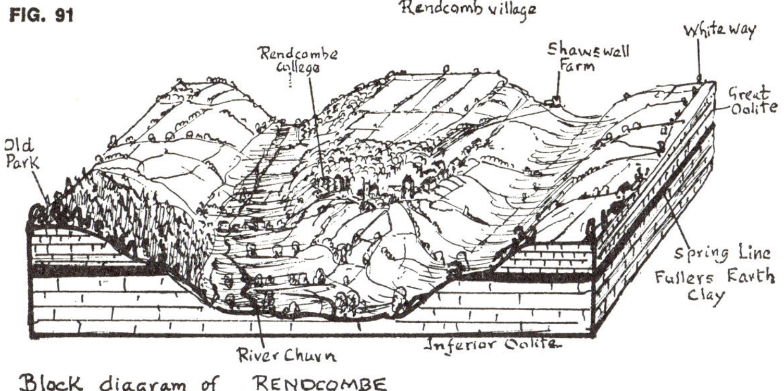

Block diagram of RENDCOMBE

THE RENDCOMB AREA

The village of Rendcomb is in a superb position perched high up on the Great Oolite on the spur of a tributary valley to the Churn. Near the village is the progressive public school, Rendcomb College, and there could hardly be a finer place or more beautiful surroundings for a school. Centuries of the history of man are written in the countryside around it, while its magnificent grounds afford opportunity for nature study on a grand scale. The block diagram at Figure 90 shows what must be one of the best cricket fields in England, surrounded by beautiful parklands leading down to an artificial lake. Round the lake *Equisetum* plants (horsetails) flourish—an excellent starting-off point for geological studies because these are descendants of the Giant Horsetails (*Calamites*) which dominated the coal forests over 280 million years ago. There is also a swimming pool, with water provided by a spring at the clay junction.

Following the Rendcomb valley to Shawswell Farm provides yet another fine example of the correlation of land utilisation and geology. The block diagram at Figure 91 shows how the Fuller's Earth clay outcrops high up on the valley slopes, outcrops which can be detected not only by the change of land use (from arable to pasture) but by the numerous small terraces caused by the clay slipping down the hill and dragging 'strings' of grass with it.

High up on the plateau on one side of the valley there is an Iron Age camp earthwork and, on the other side, a medieval salt route to Droitwich from Cirencester known as the 'White Way'. In those days of non-metalled roads, main routes deliberately avoided valleys because these became swamps in winter.

COTSWOLD SPRINGS

Many of the springs issuing from the base of the oolite are highly calcareous, so look out for the deposits of tufa. Just outside the Seven Tuns Inn at Chedworth there is a copious spring, the water from which is saturated with carbonate of lime. Here one can see mosses and small plants in the process of being petrified, while underneath is the hollow stone, or 'honeycomb' rock, which is typical tufa, a 'rock' which can be formed in a mere twenty years of geological time!

Figure 84 showed valleys which are dry at the top, but there are also valleys roundabout which are dry at the bottom even though there are springs high up on the valley sides and streams come charging down the hill. The explanation is that in many cases the floor of the valley has not yet been eroded deep enough to reach the Upper Lias clay, or perhaps not low enough to reach the level of underground water in the Inferior Oolite. The springs from the clay high up have, in fact, been drunk by the oolites of the Inferior Oolite.

Finally, notice that the Saxons preferred to site their settlements on dry land above the springs, while the Romans constructed their villas *below* so that the water would run through their baths and wash-houses.

CHAPTER 16

Cotswold Tiles and Building Stones

The typical Cotswold roof is tiled with local stone called the 'Stonesfield Slate', though it is not slate in the geological sense. To a geologist, slate is a metamorphic rock, a shale altered by pressure or folding into rocks, like Welsh slate. The correct term should be 'Tilestone', for it is, in fact, a sandy limestone which splits nicely into thin layers which are suitable for roofing. Thus an alternative name would be 'fissile limestone'.

FIG. 92

BELAS KNAP. A Neolithic Long Barrow dating from about 2000 B.C. This shows the false entrance with dry stone walls of Stonesfield Slate. This is the earliest known example of the use of the Great Oolite tilestone.

The earliest known example of its use is in the walling of the false entrance to the famous long barrow of Belas Knap, near Winchcombe (Figure 92). This dry-stone walling shows a remarkable degree of craftsmanship for the time at which the barrow was built—about 2,000 BC. True, it was repaired in 1880, but a Ministry of Works plate near the false entrance assures us that the original pattern of building was strictly adhered to during the repairs.

The Romans used the stone for roofing their villas and throughout the Middle Ages large stone tiles were similarly used. Easily dug out near the surface of the ground, they were nature's gift to builders and so these stone tiles came to be called 'presents'.

Chipping Campden is one village in which the roofs of medieval buildings are particularly well preserved. Notice the huge stone tiles at the bottom of the roofs. They often weigh 50 lb or more (a menace if fire breaks out!), and in medieval times huge oak beams were used to hold up these very weighty roofs. Their weight, in fact, is one of the main reasons for the decline of the stone tile industry, as few people today can afford the massive oak beams needed to support such a heavy load of tiles.

THE SEVENHAMPTON SLATE QUARRIES

The Great Oolite has a wide outcrop to the south-east of the Cotswolds, and Figure 93 shows how, as the dip is to the south-east, the higher or younger rocks of the Great Oolite outcrop more.

Sometimes fault block structures have preserved areas of Great Oolite, and this occurs in the old slate-quarrying area at Sevenhampton, east of Puckham Woods.

See here Figure 94, a block diagram of Sevenhampton, and Map 16 showing the fault block structure of Sevenhampton Common.

The Stonesfield Slate series of rocks form passage-beds from the Fuller's Earth clay to the Great Oolite limestones, and the slate can only be quarried in certain areas because of the underlying struc-

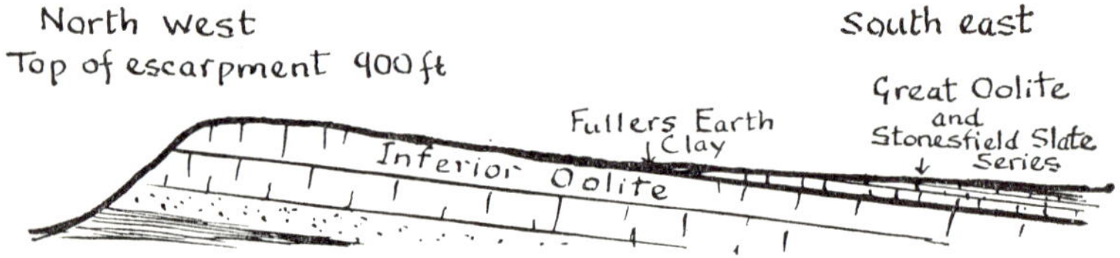

Diagram explaining the wide outcrop of the Great Oolite lying towards the south in the Mid-Cotswolds

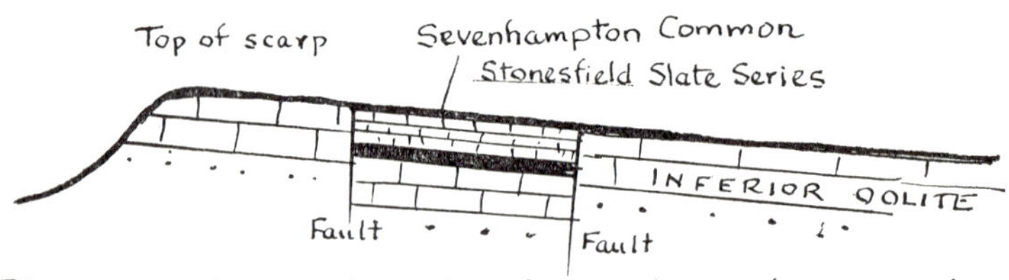

Diagram explaining how downfaulted blocks have caused Stonesfield Slate to outcrop near the scarp edge. Example – Sevenhampton Common.

FIG. 93

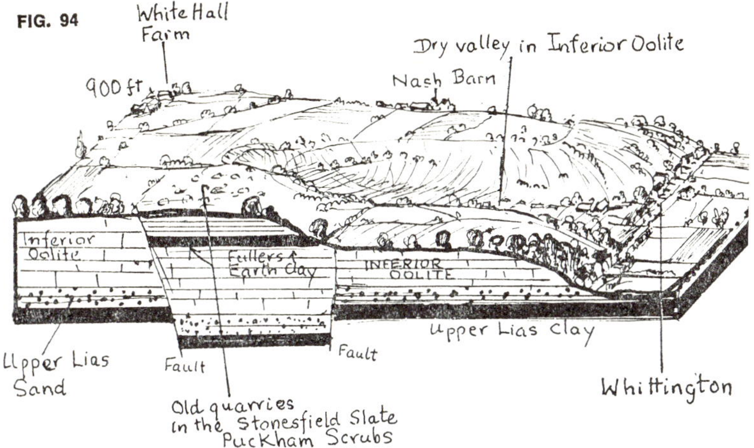

Block diagram explaining the origin of the Stonesfield Slate quarries at Puckham Scrubs, Sevenhampton Common.

tures. For example, one great structure is the Vale of Moreton anticline, where the slate series has probably been eroded off; whereas to the west, in the synclinal structures, the Stonesfield Slate series has been preserved.

Tilestones can be quarried at two horizons, at the base of the Stonesfield Slate series and also in the Chipping Norton limestone. The stone tile industry no longer exists in the Cotswolds, but in its heyday it employed hundreds of slatters who naturally preferred to quarry this sandy limestone full of tiny pieces of white mica which made it fissile and so easy to split into thin layers for tile-making.

FOSSILS IN THE SLATES

The fossils in these sandy limestones indicate that, although the rocks were laid down in the sea, the land was not far away. At that far distant time it was clothed in the rich green vegetation of the Cycads (the flowering plants had not yet come into the world), large reptiles dominated all land life, and the early mammals were still only rat-sized nocturnal mammals—probably marsupials.

Turning over stone 'slates' one can find the teeth of reptiles, casts of the fossil tree Ginkgo, and the vertebrae of reptiles such as Megalosaurus. Figure 95 shows a very rare but important find—the tiny jaw bone of a mammal. Two features which authenticate it as a mammal are its differentiated teeth and the construction of the jaw. Mammals evolved from a stem of reptiles which had differentiated teeth, and as long ago as 1818 the French geologist, Cuvier, found near Paris the remains of the first mammal in the Jurassic rocks in this series.

Apart from vertebrates, bivalves and brachiopods abound in the limestones. Most common are *Trigonia impressa*, *Ostrea acuminata* and various types of *Rhynchonellid*. Old, deserted quarries near Naunton, on the far side of Huntsman's Quarry, are the best places for fossil-collecting.

In the field, it is sometimes important to determine which is the top and which is the bottom side of a rock. This is easy enough in

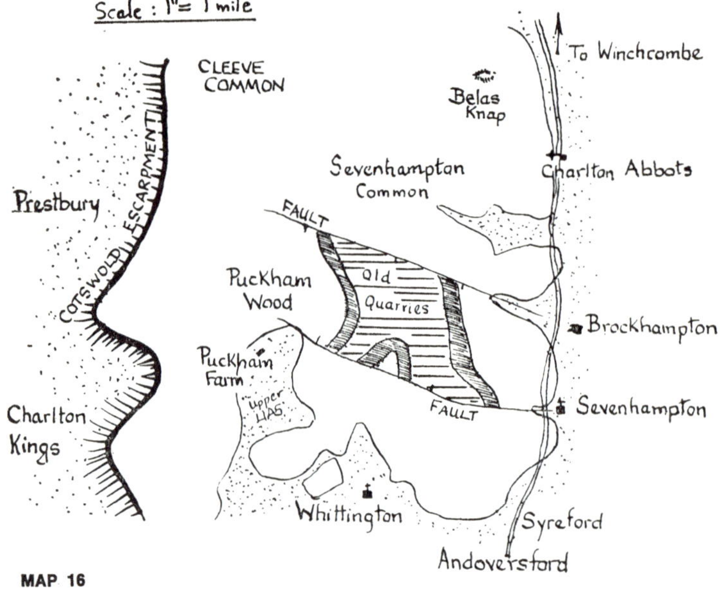

MAP 16

Geological sketch map (one inch to one mile) to show the fault block structure of Sevenhampton Common
This type of structure with W.N.W. — E.S.E faults is very common in the North and Mid-Cotswolds.
Middle & Upper Lias [ ] Inferior Oolite [ ] Fullers Earth [ ]
Great Oolite [ ]

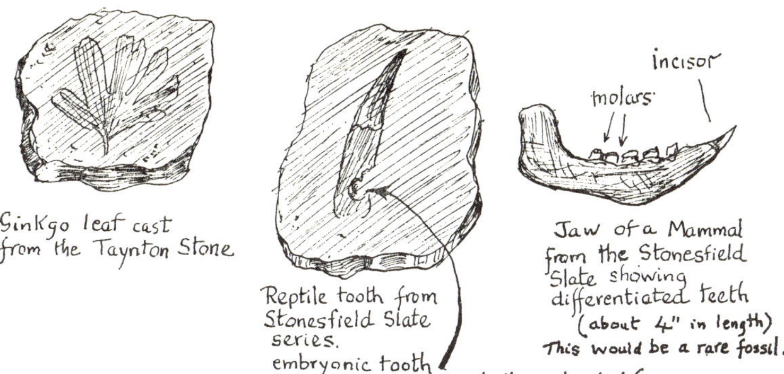

FOSSILS FROM THE STONESFIELD SLATE SERIES.   **FIG. 95**

the quarry cliff-face, as the rocks are not much disturbed from the way in which they were laid down, but when rock slabs are lying about on a quarry floor the first thing to do is to find evidence of fossil ripple marks.

Many of the rock surfaces show trace-fossils, i.e. marks made by organisms as they walked over the sand flats, or 'goblets' of sand where organisms burrowed or spewed it out. Figure 96 shows how fossils found in rocks can provide clues as to which way up the beds of rock are. Once this is known, other remains of plants and vertebrates can usually be found.

THE TECHNIQUE OF SLATE-MAKING

The various tools of the slate-making trade are shown in Figure

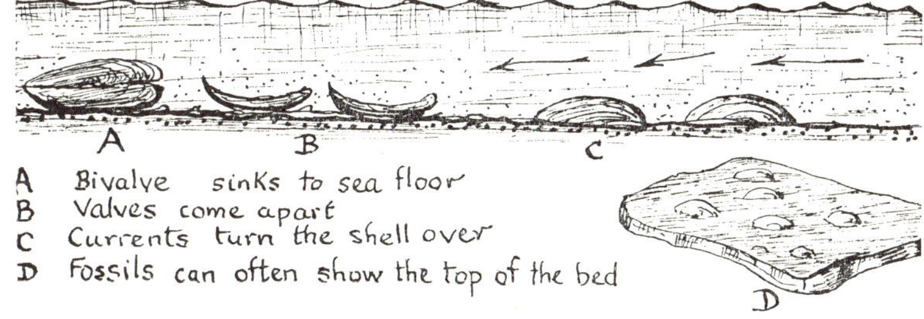

A Bivalve sinks to sea floor
B Valves come apart
C Currents turn the shell over
D Fossils can often show the top of the bed

Diagram explaining <u>one</u> of the methods used in determining the top surface of a bedding plane on a slab of rock.   **FIG. 96**

97, and all are in great demand by folk museums now that the industry is dead. Notice also the two types of slate—'presents' dug out near the surface and 'pendle', which was only sought when the supply of 'presents' became exhausted because it had to be mined far below the surface.

For pendle, slabs of rock (up to a foot in thickness) were brought up 'green', i.e. wet with underground water, and had to be covered to keep them green until the first hard frosts arrived. They were then placed out in the fields where the frost soon split them into thin tiles. If the winter was mild, much of the work involved in getting the pendle would be wasted, and for some Cotswold villages frosts assumed such importance that the church bells would be rung to summon all the men of the village to the fields when it was known that a frost was on the way.

Quite a specialised vocabulary grew up round the tile industry. The 'slatter' was the man who made the slates and the 'getter' was the man who got them out of the ground. The slates were cut into

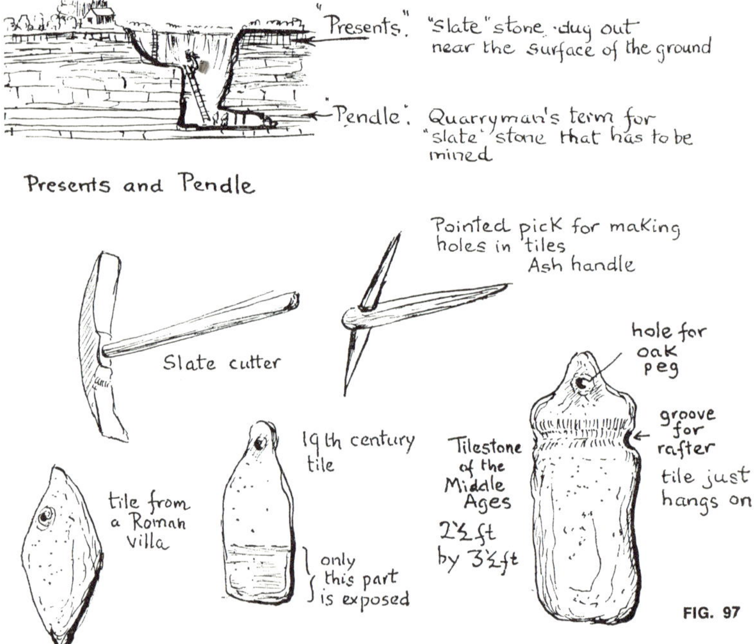

FIG. 97

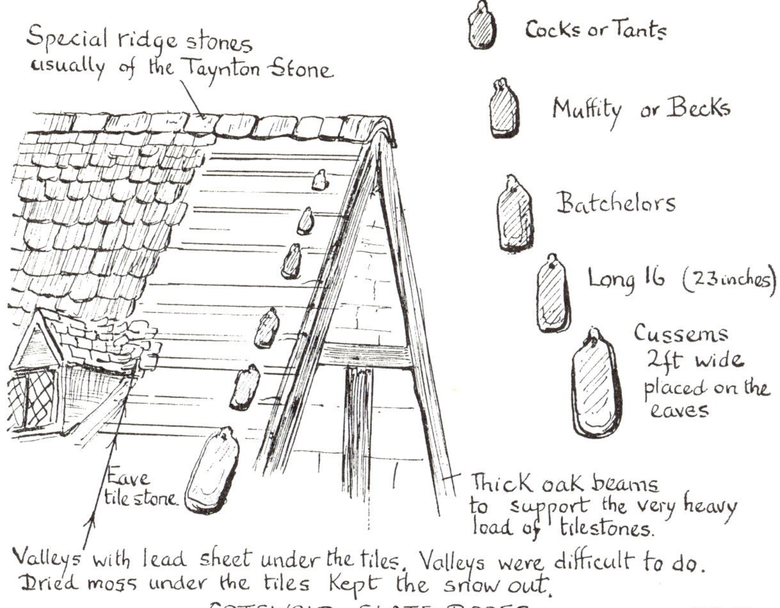

FIG. 98

varying sizes, each with a different name, the small ones being placed at the top of a roof, while the large ones, called 'cussems', went on the eaves.

Tilestones from Roman villas have been found with iron nails still in them but oak pegs were used in the Middle Ages and, later, deal. The special pick used for making the holes takes advantage of the micaceous nature of the rock, splitting off tiny layers first before the hole is pierced.

Incidentally, while many tourists stop to admire a moss-covered Cotswold roof, any slatter still alive could tell the owner that he was lessening its life by allowing the moss to remain on it. Without moss, a Cotswold roof will last hundreds of years, whereas the humic acid created by decaying moss steadily weakens and destroys slates.

BUILDING STONES

Any Cotswold village will provide examples of the use of Cotswold stone in building but for the greatest variety in types of building and for the best state of preservation, Chipping Campden is undoubtedly queen of all the pleasant places of the Cotswolds.

The raw materials of these buildings can be seen everywhere in the exposures of the dozens of deserted quarries which now riddle the Cotswolds. Labour no longer being cheap, and other building materials being readily available, only a few quarries are still being worked for building stone and the one most worth a visit is Coscombe Quarry, between Cutsdean and Stanway. It is like a vast amphitheatre surrounded by beautiful stone shelves formed as the quarrymen get the stone out from the regular bedding planes.

The stone is known as the Yellow Guiting stone, and is much seen in the Chipping Campden area and in buildings in Oxford. The greater the depth at which the stone is quarried, the deeper the yellow, but as the stone dries out it turns to light yellowish brown.

FIG. 99

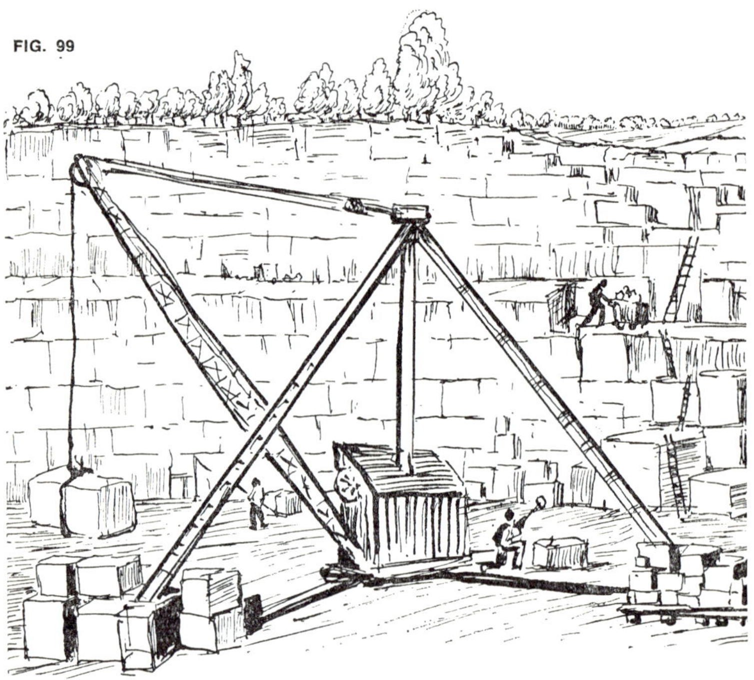

COSCOMBE QUARRY — good building stone in the lower Inferior Oolite

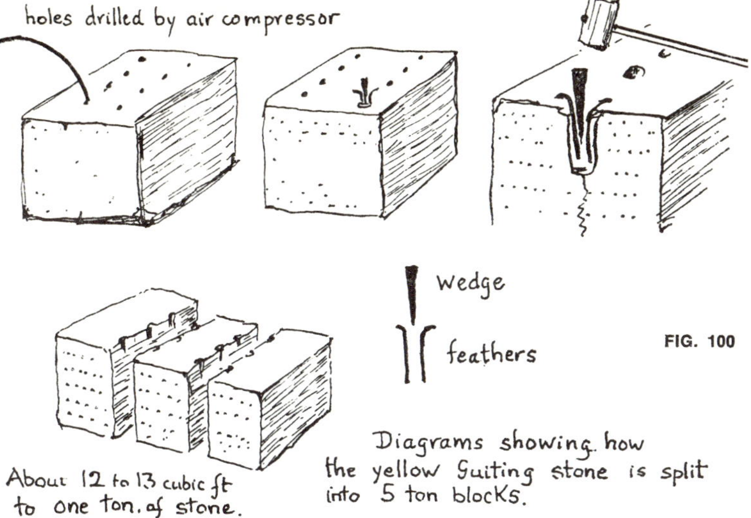

It is believed to be the same rock which, when traced southwards to Leckhampton, becomes the Pea Grit.

Figure 99, a sketch of Coscombe Quarry, shows it to be ideally sited at a point where natural processes have deposited suitable material and, by bedding and jointing, have cut it into convenient blocks for the quarrymen to handle and to market.

Figure 100 shows how the rock is split into five-ton blocks by the 'wedge and feather' method, which is cheaper and more convenient than the use of an expensive saw when only roughly-shaped blocks are required. In Westington and Farmington Quarries, however, the stone is cut into blocks by saws studded with industrial diamonds and costing over £300 apiece. This is necessary because Westington Quarry produces stone fireplaces and other ornamental masonry calling for much greater precision in the cutting.

WORKING COTSWOLD STONE

If one watches quarrymen extracting stone and masons actually working on it, it will be noticed that, most of the time, their methods are geared to the joints in the rocks. Visiting geologists are often asked about the probable structure of stone about to be worked, and analysis of the joints in some types of rock structures

FIG. 101

The old dovecote at NAUNTON.
A good example of 15th century stonework

is part of some current research being carried out at Nottingham University.

The orientation of the sets of joints in the Inferior Oolite would seem to be related to the uplift of the Jurassic rocks to form the Cotswold escarpment during the Tertiary period, some 20 million years ago—the same period as that of Alpine orogeny.

Cotswold stone, incidentally, is not at all the strong material it is popularly supposed to be, and correct quarrying and proper bedding during construction are essential to ensure that the most suitable face is exposed on completion. Careful selection is also necessary, and the irregular weathering on many exposed quarry-faces is proof of the wide range of changes which can take place within even very small natural surface areas. Natural fractures and bedding planes are also sources of deterioration unless they are dealt with in a highly skilled manner. Finally, the oolitic structure of Cotswold stone means that, although it is capable of carrying building loads, it will not withstand great pressures or blows.

CHAPTER 17

The Painswick Area

Every weekend, both summer and winter, hordes of motorists invade the Painswick area to gaze at attractive vistas and fine panoramas. At the summit of every headland of the escarpment there are sweeping views of wide areas of the Severn Valley stretching right over to the Forest of Dean beyond.

GEOLOGICAL ORIGINS

The Great Oolite is more developed here than in any other part of this particular area and, high up on the plateau, it forms a second step above the Fuller's Earth clay series, which is here about seventy feet thick.

The Inferior Oolite is not so thick as at Leckhampton (reaching about 150 feet) and part of the Lower series, the Pea Grit, attenuates from a depth of forty feet at Crickley to only a few feet near Painswick. There is, however, a marked increase in thickness of the Upper Lias sands, which reach about 100 feet and outcrop high in the valleys, affording splendid dry sites for the hill-top settlements. The Upper Lias clay is here about seventy feet thick and is, of course, the factor in ensuring the spring line at intermediate heights.

PLATFORM TERRACES

South from Stroud and Painswick, the Marlstone rock-bed forming the top of the Middle Lias is an important yet secondary topographical feature not only of the escarpment but also of the interior valleys—in which it forms platform terraces suitable for farm sites and hamlets. South of Stroud, the Marlstone forming the subedge to the Cotswolds is about a mile wide and has villages strung along it. The Middle Lias clays and shales below are about fifty feet thick.

It will be recalled that geomorphology is one aspect of 'recent'

K

geology, and is concerned with the effect of geologically 'recent' time and geologically 'recent' climate on the rocks laid down many millions of years previously. Why are the Painswick valleys so deep? Why are there so many beautiful combes in this area? Geomorphology provides us with the answer.

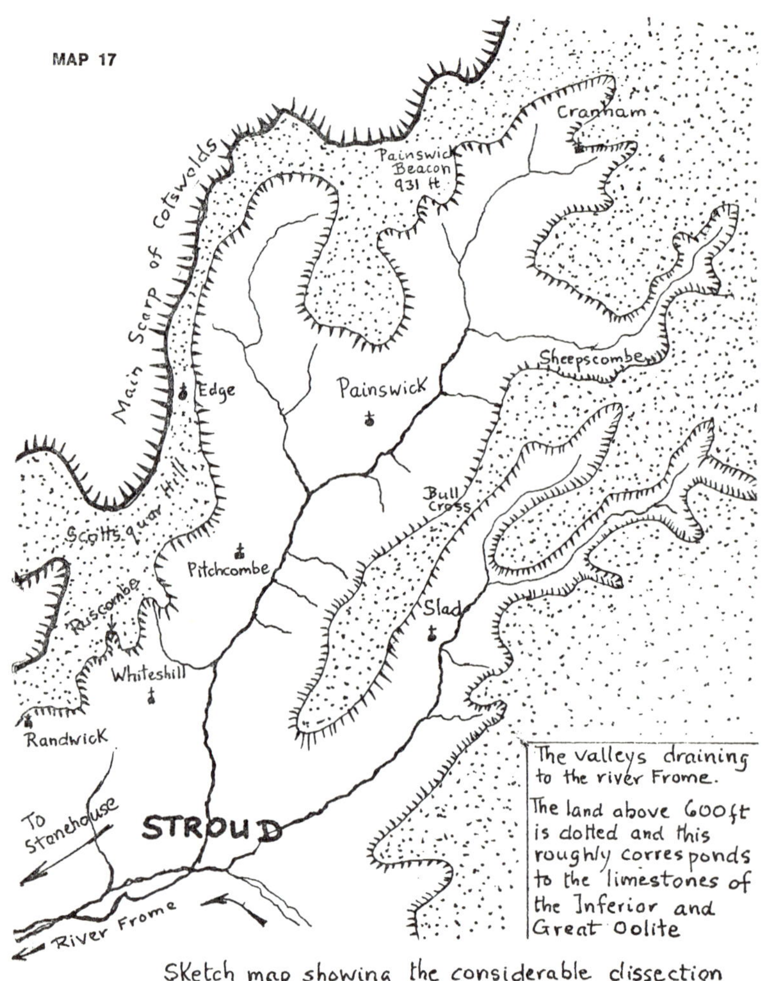

Sketch map showing the considerable dissection of the Cotswolds in the Stroud area.

FIG. 102

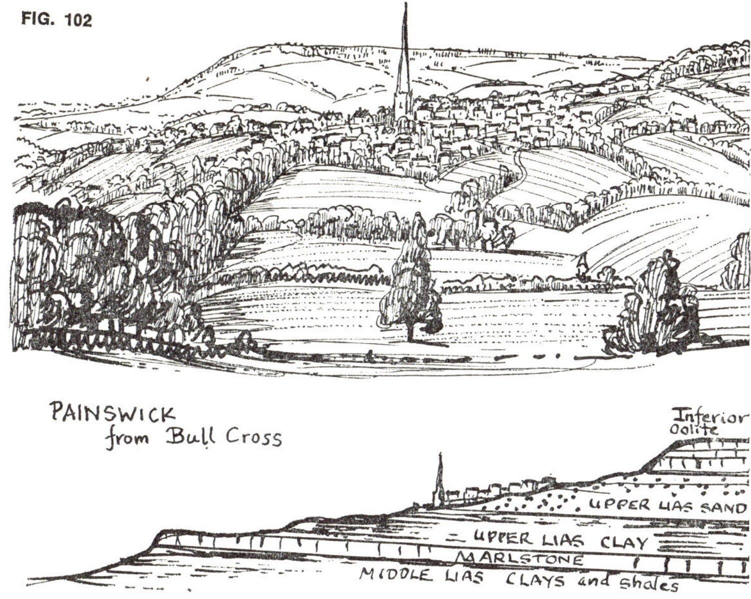

The valleys draining to the Thames are shallow, while those draining to the Severn are deeper. Obviously, then, the latter have a much swifter descent for a short distance than the streams draining to the east and they have therefore cut deeper valleys. But if we examine these valleys today we find that the actual streams are quite small (although they have provided water power) *in relation to the valleys they occupy,* and the inference is that they were most probably eroded by the much heavier rainfall which occurred during the Ice Age. In support of this theory, gravels brought down by the ancient rivers are now found some fifty feet above the present valley floors.

The main escarpment is broken at Stonehouse by the River Frome which, with its numerous tributaries, has carved out all these deep valleys. These valleys tend to be at right angles to the main valley and, upstream, terminate in deep coombes which are themselves at right angles to the valleys. There is no doubt that, originally, rivers worked their way along weaknesses in rocks (e.g. valleys tend to be eroded in clays). In this area there are north/south faults and lateral faults, and these have apparently initiated the drainage system. The faults often cause displacements in the strata, so the Marlstone on one side of the valley is at a different level from the other side, which also accounts for the varying depths of the valleys.

THE MAIN BEAUTY SPOTS

The sketch of Painswick does not do justice to the beauty of the site of this delightful little Cotswold town. No picture taken from any angle could do that, for Painswick is sited high and dry on a promontory between two deep valleys. Inevitably, the attractiveness of the town and its superb site have brought traffic problems, especially during the holiday season, for the town has only been able to grow along the spur on which it stands and there is,

FIG. 103

The SLAD valley, near Bull cross.
   The farm is on Upper Lias Sands, above the spring line

therefore, only one route through it.

The Slad valley (Figure 103), only a few miles from Painswick, achieved international fame through Laurie Lee's account of his boyhood in the village of Slad, and now *Cider with Rosie* is even a set 'Eng. Lit.' textbook in the United States! The sketch shows the great depth of the valley, which ends in the usual coombe, an erosion feature perhaps not entirely due to past climates in this instance but also associated with camber and slip in the rocks.

FIG. 104

Many of the old mills in this and nearby valleys are now private houses and the most interesting from the geological angle is Rock Mill. Here the stream has cascaded over the Marlstone, which can be seen outcropping just over the Mill House.

Haresfield Beacon is one of Gloucestershire's most popular beauty spots at all times of the year. It is also a vantage point from which one can clearly see the root cause of the Severn Bore—the wide meanders of the river which form the 'funnel' and the long 'stem' which leads to Gloucester. From here, too, can be seen in the far distance the graceful span of the new Severn Bridge.

But, above all, Haresfield Beacon is a geologist's dream spot! Slip, camber and other fractures and distortions obscure the true relationships of the strata all over the Cotswolds but on Haresfield Beacon there is a geologist's classic exposure showing the exact junction of the Inferior Oolite and the Upper Lias sands (see Figure 106). Every amateur geologist should make a point of finding this

FIG. 105

Haresfield Beacon

HARESFIELD BEACON

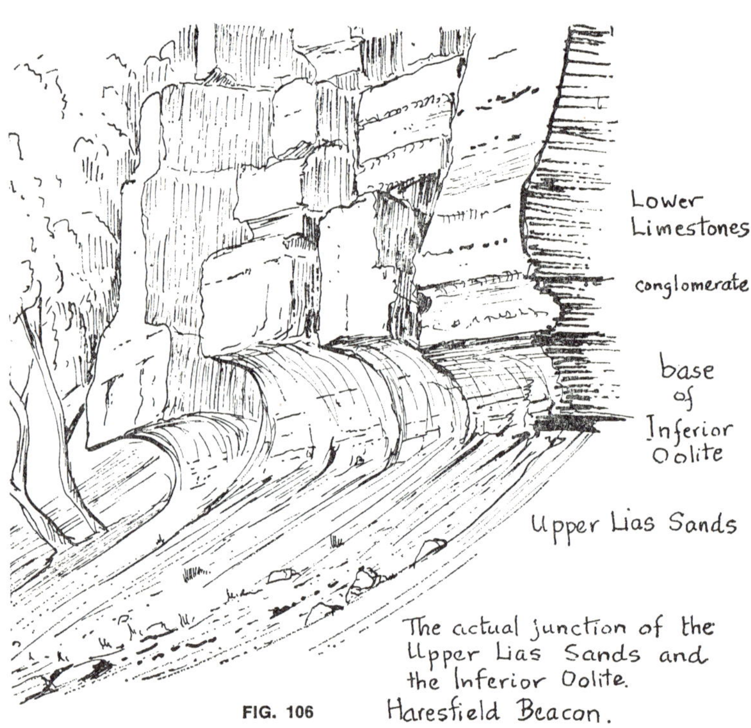

The actual junction of the
Upper Lias Sands and
the Inferior Oolite.
Haresfield Beacon.

FIG. 106

exposure which is just below the Ordnance Survey Trig. point pedestal, but easy to miss on a casual walk.

The villages of Westrip, Randwick, Ruscombe and Selsey are all situated on the Upper Lias sands and the alignment of the houses is contoured round the coombes. At night, the lights in the houses conveniently sketch out the contours of the landscape for the viewer! As Stroud expands, housing tends to retreat around the valley contours. There is little need here for town planning, as geology determines what can and cannot be done.

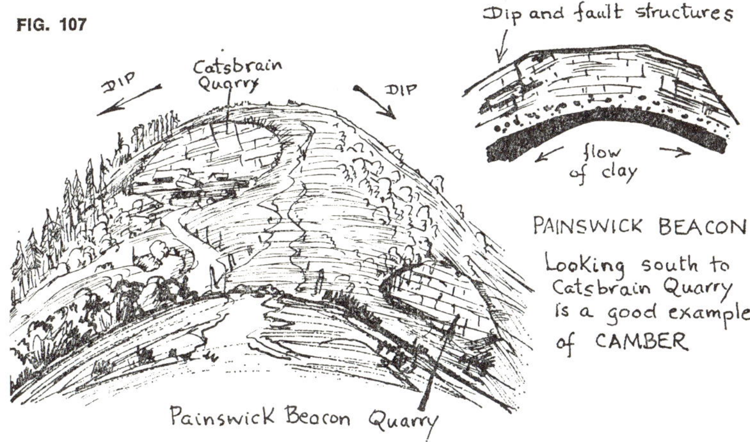

Painswick Beacon (Figure 107) is another favourite beauty spot in this area. At a height of 931 feet, it has commanding views right across the Severn Valley to Robin's Wood and May Hills and beyond. This also is the site of a former Iron Age hill-fort, being conveniently situated on a long, narrow ridge of limestone, and because the ridge is so narrow there is much cambering on both sides. The best view of this can be obtained from the Beacon summit looking south to Catsbrain Quarry on the left and the Beacon quarry on the right, each with dips towards the valley.

Looking at this, the beginner in geology would probably immediately exclaim, 'A nice anticlinal structure!' It is not, of course, and the diagram and sketch in Figure 107 show the complete camber as seen in one view.

CHAPTER 18

The Northern Malverns

The Malvern Hills form a north/south range about seven-and-a-half miles long, rising like a wall from the south-western approaches to the Midlands.

The hills are high because the rocks are hard and crystalline and have resisted erosion, and also because they have been heaved violently upwards from the depths of the earth's crust. They form a marked boundary between two distinct regions.

On the east, there are the populated plains of the Midlands with vast stretches of arable land but with new industries and new towns springing up everywhere. On the west, it is a region of wooded scarps and vales, with rather a sparse population, the Welsh borderlands of Herefordshire.

On the east, the rocks are the flat Triassic rocks laid down in desert conditions some 250,000,000 years ago. On the west, there are much older rocks which are also folded so that the resultant scenery is far more picturesque.

Viewed from the Midlands side at about ten miles distance, the Malvern range can be seen to be much higher in the north than in the south, and the generally accepted explanation of its present-day appearance is that this upthrust block of ancient rocks has been pushed up with a greater thrust in the north than the south.

Figure 108 shows the sudden rise of the Northern Malverns by North Hill, while Figure 109 shows that, at Chase End Hill in the extreme south, the rise from the plain is not so imposing.

The rocks of the Malverns are particularly interesting because they are among the most ancient in Britain, and known to geologists as the Pre-Cambrian rocks. Some recent research by Dr D. C. Rex of the Department of Geology and Mineralogy at the University of Oxford has established that the period of their crystallisation from sedimentaries into rocks occurred between 580 and 600 million years ago. Previously, about 1,000 million years ago, they may have been very ancient shales or sandstones. The method used by

FIG. 108

Dr Rex—known as the isotopic age determination of rocks—is based on the decay rate of the radio activity in certain minerals found in crystalline rocks. In this instance, hornblende, biotite and muscovite were the three minerals selected and submitted to the potassium-argon technique.

FIG. 109

The southern end of the Malverns. Chase End Hill, 625 ft. This shows the contrast with the northern end. There has been more upward thrust in the north.

THE ROCKS OF THE MALVERNS

The Malverns are clues to the nature of the deepest parts of the earth's crust—conveniently thrust up in this particular area for the geologist to look at. It would seem that the earth's crust is intensely folded at its greatest depths and the rocks there are harder because of their crystalline texture.

There are numerous quarries round North Hill in all of which the rocks are crystalline and the observer does not need to be a specialist in mineralogy to recognise the more common minerals, all silicon compounds of various kinds. As a brief guide to their identification:

Feldspar minerals are usually white or pink—pink feldspar gives the rocks a reddish tinge.

The harder vitreous or glass-like mineral is quartz.

Dark green to black mineral is hornblende.

The mica minerals can be of two kinds, a clear mica called muscovite and a dark iron mica called biotite.

Biotite and hornblende are easily confused. The former is black, shiny and flaking but the hornblende does not flake, although often of the same colour.

Rocks can be identified by texture and structure as well as by colour. Stand back and survey each quarry-face as a whole to get an idea of structure. In some quarry-faces it will be seen that what at first appear to be gigantic bedding planes are not bedding planes at all but lozenge-shaped masses of rock dovetailing into one another. Such a typical cliff-face is shown in Figure 110 and is known as a 'schist', which means a crystalline metamorphic rock that foliates. It often breaks in a wavy uneven surface and this property is called 'schistosity'. Schists are named after their characteristic mineral, which is hornblende in the Malverns, so this rock is a hornblende schist.

Schists are produced under intense pressure and, looking at the quarries in the Malverns, even a layman can sense that some enormous lateral force must have pushed the rocks into those irregular layers.

Dynamic pressure has, in fact, changed rocks which were formerly sedimentary rocks into schists, and when this change has been brought about by pressure, or by heat from injected molten rock, or by both phenomena, the resultant rocks are known as 'metamorphic'.

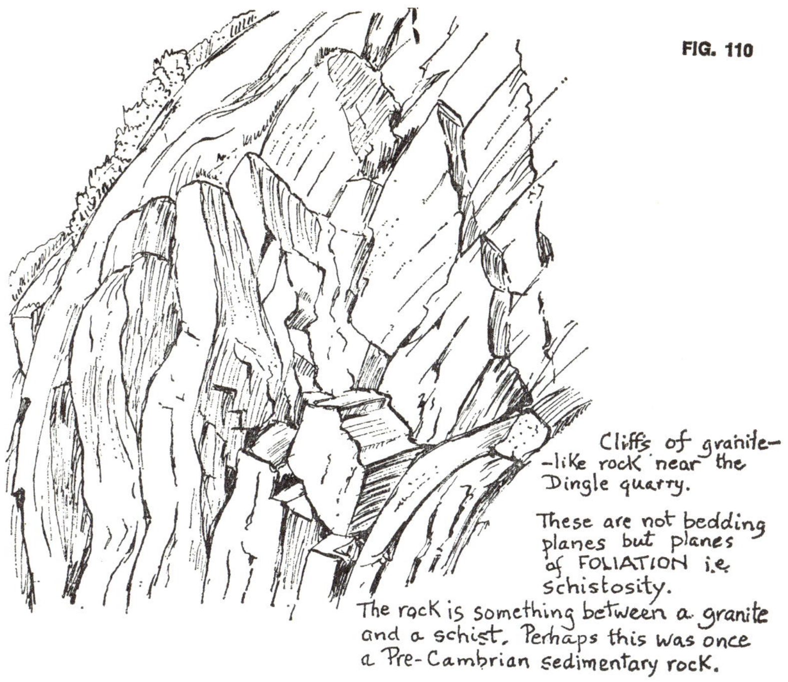

This kind of rock, which breaks into irregular layers and lumps, is really only suitable for road metal, and the quarries still busily working in the Malverns are mainly supplying material for the new motorways. It is excellent material for this purpose because the rock is so hard that tungsten carbide steel tools have to be used for drilling holes into the rock before blasting.

Some old houses in the area have been built from Malvern rock but are merely further proof that this is not a good building stone because so much cement has been needed to bond the irregular rocks. Probably the real attraction in each instance has been the ornamental character of the stone and the fascination of its many changes of colour.

Particularly attractive are the great white veins of some injected material in these rocks. Closer inspection of these beautiful streaks of mineral in the quarries reveals that they are feldspar and quartz, thus making a puzzle within a puzzle for the geologist.

The original sedimentary rocks were apparently metamorphosed by pressure and then further metamorphosed by the injection of molten material. Or did it happen the other way round? Even today geologists argue endlessly about the Malverns! There is no general agreement about the exact sequence of events in the evolution of these most ancient rocks to their present state.

Rocks which were once molten and have now crystallised out are called igneous rocks, and can generally be recognised by the absence of well-defined bedding planes and by their particular texture rather than by their colour.

FIG. 111

GREEN VALLEY AND IVY SCAR

After the quarries of North Hill come the slopes of Worcestershire Beacon and probably the most spectacular scenery of all the Malvern range—Green Valley by St Ann's Well.

There is no doubt that this scenery, which is so like Dartmoor, is related to the granite-like type of rock which probably forms the main mass of the Northern Malverns. In places, it is classified as a quartz diorite or granodiorite and this whaleback type of relief is usually associated with the weathering of granite hills.

Green Valley is like a coombe and many similar features can be found on the eastern and north-eastern sides of the range. These coombes may have been eroded by small glaciers during the last glacial period.

FIG. 112

A well-defined public footpath leads out of Green Valley to the famous Ivy Scar rocks which tower over Malvern Town, whose majestic Priory Church is almost like a human echo of the dark rocks erupting out of the smooth granite hills.

This dark, heavy rock is classified as a micro-diorite, a kind of dolerite which is really an igneous intrusion into the older granite schists—and the gneisses which also occur in the Malverns.

Gneiss is difficult to define or describe because it is so varied but in general it is a coarse-textured metamorphic rock with the minerals in parallel streaks but lacking the characteristic property of schists to break in a wavy, uneven surface.

The reason geologists believe that the Ivy Scar rocks are a later intrusion of molten rock is that they are so 'fresh' looking, i.e. there is not much alteration of the constituent minerals. There is also the fact that from one viewpoint can be seen the various flow patterns and bedded structures of the molten rock as it was squeezed in or injected from the depths of the earth.

From this point there is a climb of over 1,000 feet over Sugar Loaf Hill and then the path leads to the Dingle in West Malvern where there is a spectacular view across to the wooded limestone scars of the Silurian rocks which will be discussed later in this chapter.

Dingle Old Quarry is well worth a visit for here is further evidence of later intrusions of molten rock. On the north side of the quarry can be seen the actual zone of contact of a later injected rock, new rock injected into the older quartz and feldspar rock. It looks as if it was injected along one of the great thrust planes which are so common in the Malverns and, according to some opinions, these thrust planes are involved in the emplacement of the hills themselves.

Also worth a visit are the old deserted quarries close to Upper Wyche, particularly the large one known as Earnslaw Quarry, as these give a valuable insight into the real nature of the Malvern rocks. Here can be seen the immense cleavage planes stretching up to the top of the cliff and even the direction from which the pressure came to give these rocks 'the great squeeze'.

Great ribs of granite-like intrusions stand out like pinkish-white dykes. This suggests that these feldspar quartz intrusions may have had something to do with the conversion of the original metamorphosed rocks into granites and grano-diorites.

MINERAL CHANGES

Examining the rocks in all these quarries, the most striking feature which will be noticed is the variety of colours, particularly greenish minerals. Remember that rocks are merely aggregates of minerals and that in very ancient rocks the minerals decompose and change into other minerals—especially when they are subjected to heat and pressure or saturated by underground fluids. A common green decomposition product seen in Malvern rocks is chlorite and another is epidote.

The black, shiny iron mica called biotite often occurs in thick veins and this mineral, together with the other kind of mica, white mica called muscovite, has been formed by the so-called 'granitisation' of the Pre-Cambrian rocks. Granitisation is a process in which some form of igneous material or another invades sedimentary or metamorphic rock, producing mixtures which eventually alter the rock so that in texture and composition it becomes granite.

Epidote is a complex silicate of calcium and aluminium with water which forms in most metamorphic rocks, following cracks and seams, but in many Malvern quarries it has formed thin green crusts.

Some lumps of rock are dark and heavy because they consist almost entirely of a mass of hornblende crystals, hornblende being a complex hydrous silicate containing calcium, magnesium and iron. In many types of rock the hornblende has changed into biotite and in so doing given rise to the secondary minerals, chlorite and epidote.

Finally, some of the rocks show an irregular layered arrangement in which the minerals are in parallel streaks or bands but do not show cleavage, i.e. schistosity. This type of rock could have originated either from a granite which has been subjected to enormous pressures or from an original sedimentary rock which has been metamorphosed into a schist and then later changed into a kind of granite gneiss by igneous permeations. Confusion built upon confusion!—which is what the great Malvern debate will be for many years to come.

THE SPRINGS

Seeing so many great fissures, it is natural for the observer to assume that the surface rainfall fills them up with water underground. Certainly there are many springs in the hills but most frequently they occur where the crystalline rocks come close to the junction of sedimentary rocks.

The claim is made that this water is absolutely pure and the well-known Malvern water obtained from Holy Well near Malvern Wells is even bottled and sold to the public as Malvern Water, claiming no special property other than this purity, an ironic comment on the state of mains water in our overcrowded island!

THE SILURIAN ROCKS

On the western side of the Malverns, for a distance of over five miles, a series of rocks outcrops, forming ridges of sandstone or limestone or valleys in the softer rock called shale. This series is

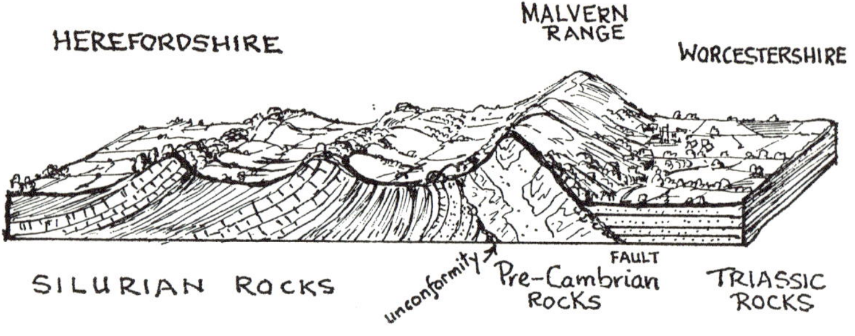

FIG. 113   THE MALVERNS — looking north.

known as the Silurian system as these sedimentary rocks were laid down in the Silurian sea about 440 million years ago when the Malverns (which were then a much-eroded ancient mountain chain) were submerged beneath the sea.

Figure 113 shows the great contrast between the flat Mesozoic rocks (the Triassic system) on the east and the folded Silurian rocks on the west.

Looking down on these wooded limestone scarps and the open field of the valley floors, the observer is provided with a 'text book' example of how to recognise limestones and shales forming escarpments and valleys respectively.

To look at this Malvern scenery is not only a lesson in geology but also a practical lesson in geography affording a real understanding of some fundamental principles of all scenery and never to be

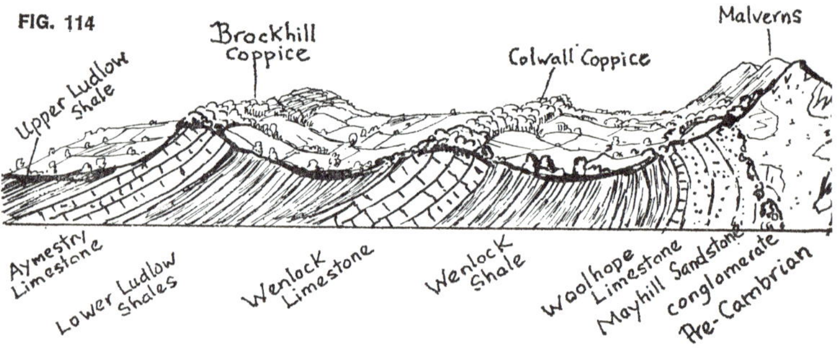

THE WOODED LIMESTONE SCARPS IN WEST MALVERN

forgotten because it was something actually seen.

There are even small gaps in the scarps formed by the Aymestry limestone so that an afternoon spent on this part of the Malverns will give any geography student the key to an understanding of the physical build of much of south-eastern England and even the Paris Basin in France.

Between British Camp and the Wyche railway tunnel the observer can follow a line of wooded scarps represented by the two dominant limestones of the Silurian system, the Wenlock and Aymestry limestones. It is obvious from the block diagram in Figure 114 that the younger rocks outcrop towards the west.

FIG. 115

The alignment of the Malvern Range
view northwards to Worcestershire Beacon

There is an abstract beauty in the patterns of the crystalline rocks of the Pre-Cambrian series but the Silurian rocks offer for most visitors to the Malverns a livelier interest because they are so rich in fossils. Corals, brachiopods and trilobites are to be found, the best places for the hunt being between Brockhill Coppice and Colwall Coppice in the old limestone quarries.

From above Colwall there can be seen in the hill immediately above Gardener's Common a clue to the presence of faults because there is a sudden change in direction of the limestone scarps. In fact, nature here supplies a neat geological map by covering the limestone outcrops with woods! In this hill a fault has caused the Wenlock limestone to shift its position in relation to its distance from the Malvern range.

L

FIG. 116

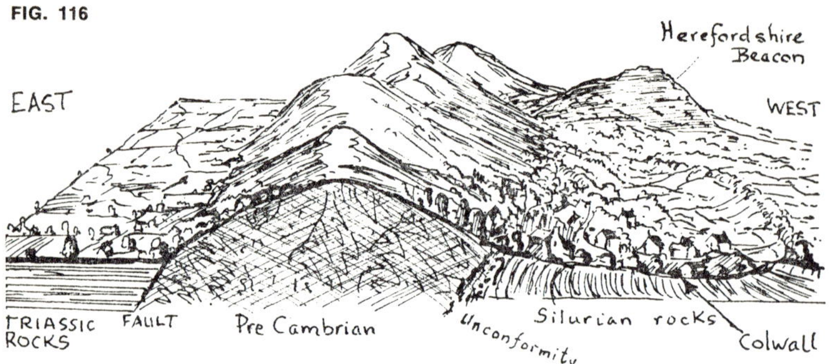

The Malvern Range looking south by the railway tunnel – Upper Wyche. It shows how Herefordshire Beacon is out of alignment. A lateral force has pushed it half a mile to the west

The next outstanding feature is Herefordshire Beacon, but before going on to study this the rambler should pause to notice the following—that looking north here the great alignment of the Pre-Cambrian hills shows up very clearly (see Figure 115) and looking south from the viewpoint shown in Figure 116 it can be seen that Herefordshire Beacon is out of alignment. Some great force has pushed it half-a-mile to the west.

HEREFORDSHIRE BEACON

Although not the highest hill in the range, Herefordshire Beacon is rather dominant because of its peculiar offset position and no doubt this is the reason Iron Age tribes chose it for a hill-top fort. The great earthworks accentuate the flat-topped nature of this hill which is really high enough to be called a mountain.

This mountain is made of a tough crystalline rock—hornblende gneiss—and one may marvel at the primitive Iron Age tribes who managed to excavate immense ramparts in such hard rock when their tools were merely stone axes and bone shovels.

The answer, of course, is glaciation. Within the last million years the ice advanced from Wales into Worcestershire and intensive frosts lasting thousands of years heaved up rocks and, by a freeze-thaw action, split them into smaller jagged fragments. This sub-soil is found all over the hills, except where the rock has been scoured clean to reveal the marks left on their surface by the transporting ice. There is also the fact that, even with stone tools, it is possible to hew out slabs of hard rock where cleavage planes occur.

Figure 117 shows what are believed to be the main structures of the Beacon.

The main Malvern range has a ridge-cum-saddleback shape but Herefordshire Beacon has a flat-topped appearance with a very steep side to the west.

The group of hills to the east known as Broad Down and Tinkers Hill are composed of Pre-Cambrian rocks and are, of course, part of the Malverns, but these particular rocks are different from the other Pre-Cambrian outcrops. They consist of very ancient volcanic rocks (rhyolites and spilites) which are very much altered and resemble the Pre-Cambrian rocks of Shropshire, called Uriconian rocks.

The Pre-Cambrian period may cover a period of 2,500 million years but it is impossible to classify Pre-Cambrian rocks by fossil content because there are only a few traces.

Classification is therefore done by lithology, the science of rocks as mineral masses, e.g.

(a) Sedimentary, e.g. Longmynd rock of Shropshire.
(b) Metamorphic—schists, gneisses.
(c) Volcanic—a series best seen at Caer Caradoc in Shropshire. The rocks of Broad Down and Tinkers Hill are Pre-Cambrian volcanic rocks and resemble the Uriconian volcanics of

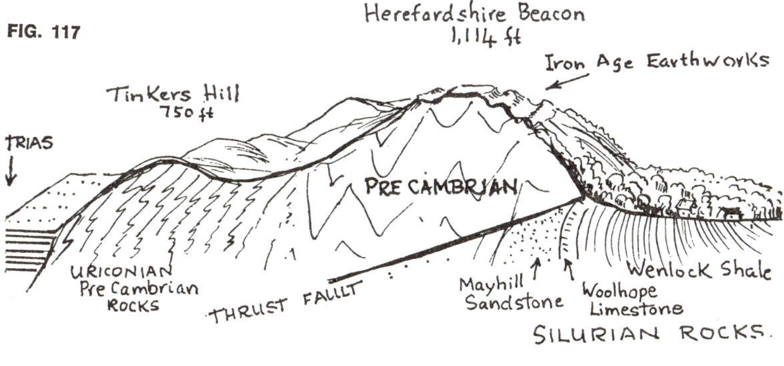

THE GREAT THRUST THAT HAS PUSHED HEREFORDSHIRE BEACON HALF A MILE TO THE WEST. It occurred before the Triassic rocks were laid down. (about 290 million years ago)

Shropshire. The geological survey calls the Malvern volcanics the Warren House series.

Uriconian rocks are probably younger than the main mass of Malvernian rocks. Figure 117 is based on the assumption that the Herefordshire Beacon Hill mass has been thrust half-a-mile to the west to reveal this mass of Uriconian rocks.

Taking advantage of the hard impervious nature of these rocks, a small reservoir has been constructed at the foot of Broad Down. This adds to the beauty of the scenery and the locals call this the Lake district of the Malverns.

CLUTTERS CAVE

A short distance south from the British Camp the footpath passes by a medieval shepherd's cave called Clutters Cave and this is one of the best places at which to study in detail the various flow structures in the Uriconian volcanic rocks.

Every week-end crowds of tourists are attracted to Herefordshire Beacon by its outstanding views, and here again nature has drawn for the observer a neat geological map.

FIG. 118

Clutters Cave. (near Herefordshire Beacon)
A man-made cave in the Uriconian rocks.

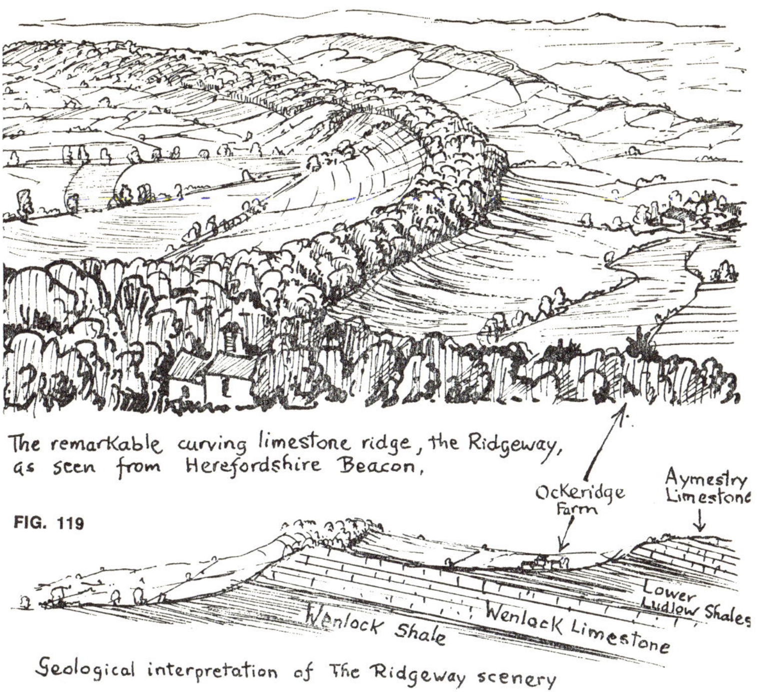

Geological interpretation of The Ridgeway scenery

Spread out on the west is a remarkable wooded ridgeway curving round from the foothills into the distant landscape. This ridgeway is the Wenlock limestone and the vale to the west of this is the outcrop of the softer Wenlock shale. A repeated scarp to the east denotes the presence of the Aymestry limestone (see Figure 119).

All students of geography and geology in schools, colleges and universities should spend at least half-an-hour on the western side of Herefordshire Beacon reading the book so neatly set out by nature.

CHAPTER 19

The Southern Malverns

The midpoint on the Malvern range is on the hill north of Herefordshire Beacon and just opposite Little Malvern, but for the purpose of this chapter the region covered is south of the Beacon, where not only are the hills lower but the scenery on the western side begins to change.

Follow the usual ramblers' summit-of-the-ridge footpath. This always seems to follow the Red Earl's Ditch, a medieval boundary earthwork separating the counties of Worcestershire and Herefordshire. The frequently-outcropping hard crystalline rocks of the Pre-Cambrian show the usual variations from quartz feldspar rock to gneiss and schist.

The dominating man-made feature which looms up is the Obelisk, erected in 1812 to Lord Somers, Baron of Evesham, and to the Cocks family. It would be impossible to build such a monument in Pre-Cambrian rocks, so oolitic limestone was dragged all the way from the Cotswolds to build this obelisk. Geology students are sometimes puzzled when they find fragments of Jurassic oolite in the nearby gully, not realising that the workers of 1812 dropped pieces of oolite on the way up the hill!

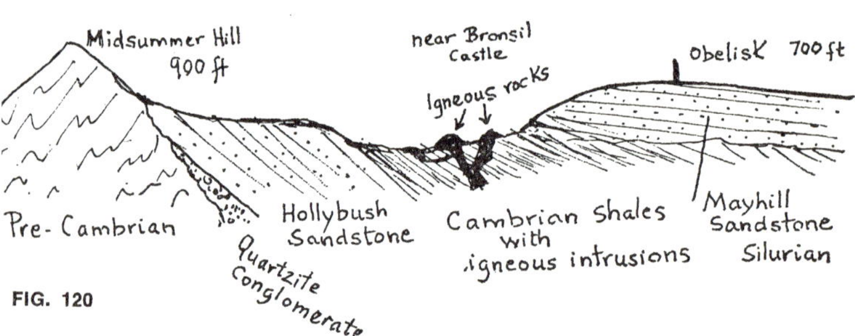

FIG. 120

View looking south to the west of Midsummer Hill. Here, the Cambrian rocks dip off the Malverns. They consist of basal conglomerate, Hollybush Sandstone and shales.

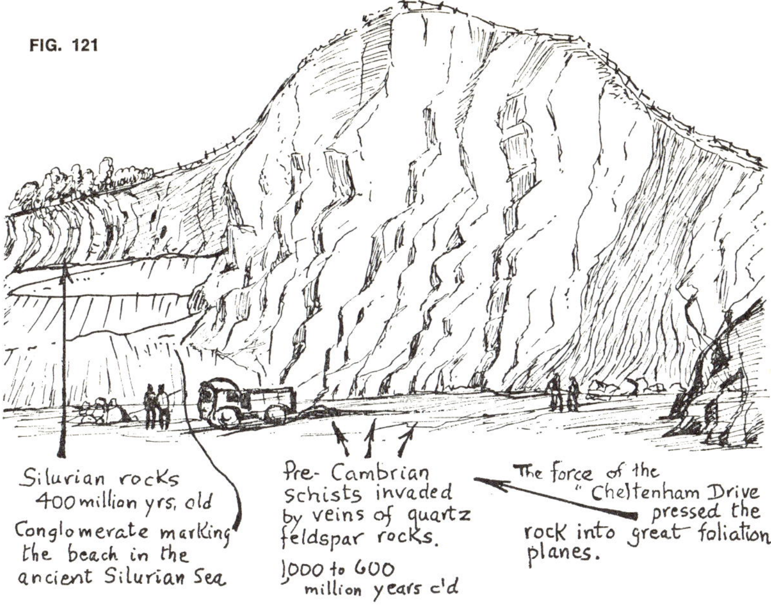

THE GULLET QUARRY

The Obelisk stands on a high, sloping plateau forming a marked topographical feature on the western side of the Southern Malverns. This plateau is the May Hill sandstone, a Silurian sandstone estimated to be 1,100 feet thick. These arenaceous rocks are often referred to as the Llandovery series. Figure 120 gives an explanation of the scenery but scientific purists should note that the geological section has been 'wangled' very slightly just to show how Cambrian shales form valleys further south!

THE SILURIAN PASS

Between Hangman's Hill and Swinyard Hill is a pass from Pink Cottage to St Malm's Well—not a very low pass, merely a col. This is an important area of study for geologists because faults have let down masses of Silurian Mayhill sandstone into the Pre-Cambrian rocks and the crystalline rocks have been covered. Some geologists believe that this is evidence that the Malvern range was once submerged in the Silurian sea.

THE GULLET QUARRY

This place is a 'must' for all visitors. Looking up at the quarry face it is tremendously impressive to see the enormous planes of schistosity stretching right up to the cliff-top. The forces which thrust these hard rocks into one great squeeze movement must have been overwhelmingly powerful. Yet it must not be supposed that this was an explosive and catastrophic movement. The forces were acting slowly, although with irresistible power.

Figure 121 shows the great thrust planes and just as important is the great unconformity. This is shown in greater detail in Figure 122.

This good section was revealed by the quarry company in 1965 and it shows sedimentary rocks of Silurian age dipping steeply away from the Pre-Cambrian schists. They are of the Llandovery series and consist of fine-grained sandstones and siltstones with many fossils of the coral and brachiopod type. Curiously enough, there are no shells to be seen, merely casts, but this is because percolating water has decalcified them, i.e. dissolved out the hard parts leaving only casts of the fossils.

FIG. 122

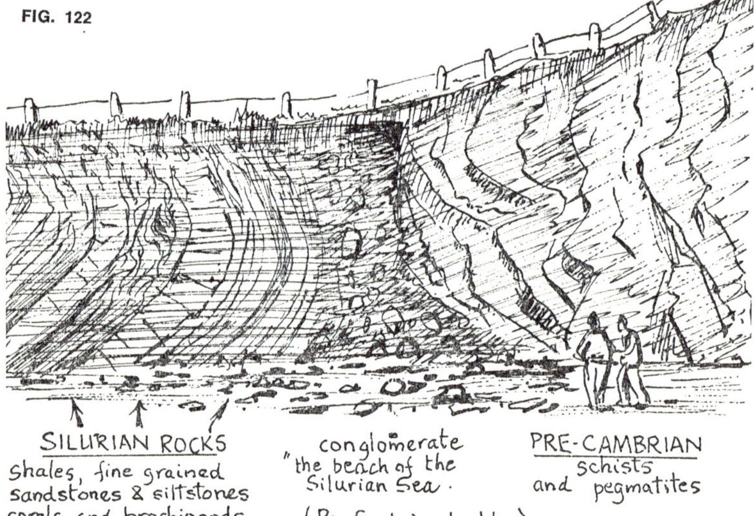

The famous unconformity in the GULLET QUARRY in 1965

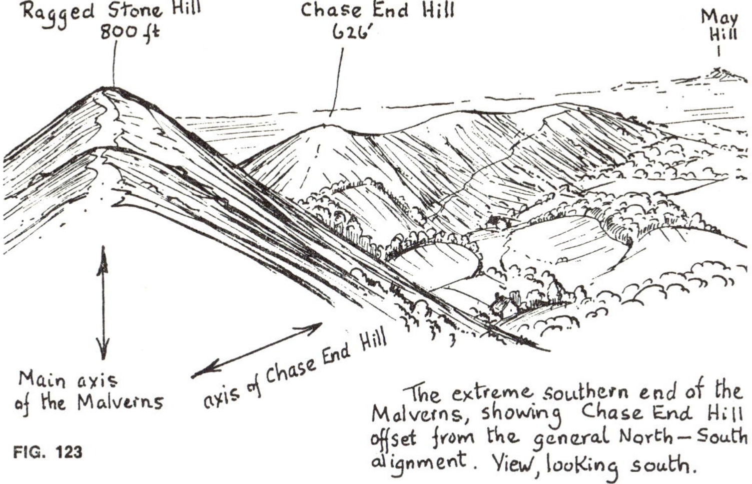

FIG. 123

Right at the junction with the Pre-Cambrian rocks is a conglomerate which consists of large boulders and pebbles of Pre-Cambrian rocks, an actual beach by the Silurian sea some 440 million years ago. The writer has made application to the Malvern Conservators for the possible wiring in of this exciting rock section so that it can be preserved for all time as a contribution to scientific understanding.

MIDSUMMER HILL

This is another typical high-ridge type of hill of gneisses and schists with many white outcrops of hard, milky, quartz rocks. It is also the site of an Iron Age earthwork camp.

Following the Red Earl's Dyke again, the rambler looks across to Ragged Stone Hill and it is surprising to see a wooded coombe tucked right into the hill. Here there is evidence in small outcrops and quarries that sandstone rocks have been let down by faulting into the Pre-Cambrian. This sandstone is of Cambrian age and is known here as the Hollybush sandstone.

Many visitors who go to the Malverns 'do' the entire range by walking the whole length of seven-and-a-half miles (actually about ten miles with the ups and downs!). The path to the south gets lower and lower and Chase End Hill is the final goal at 626 feet. The 'start' at North Hill is 1,100 feet high.

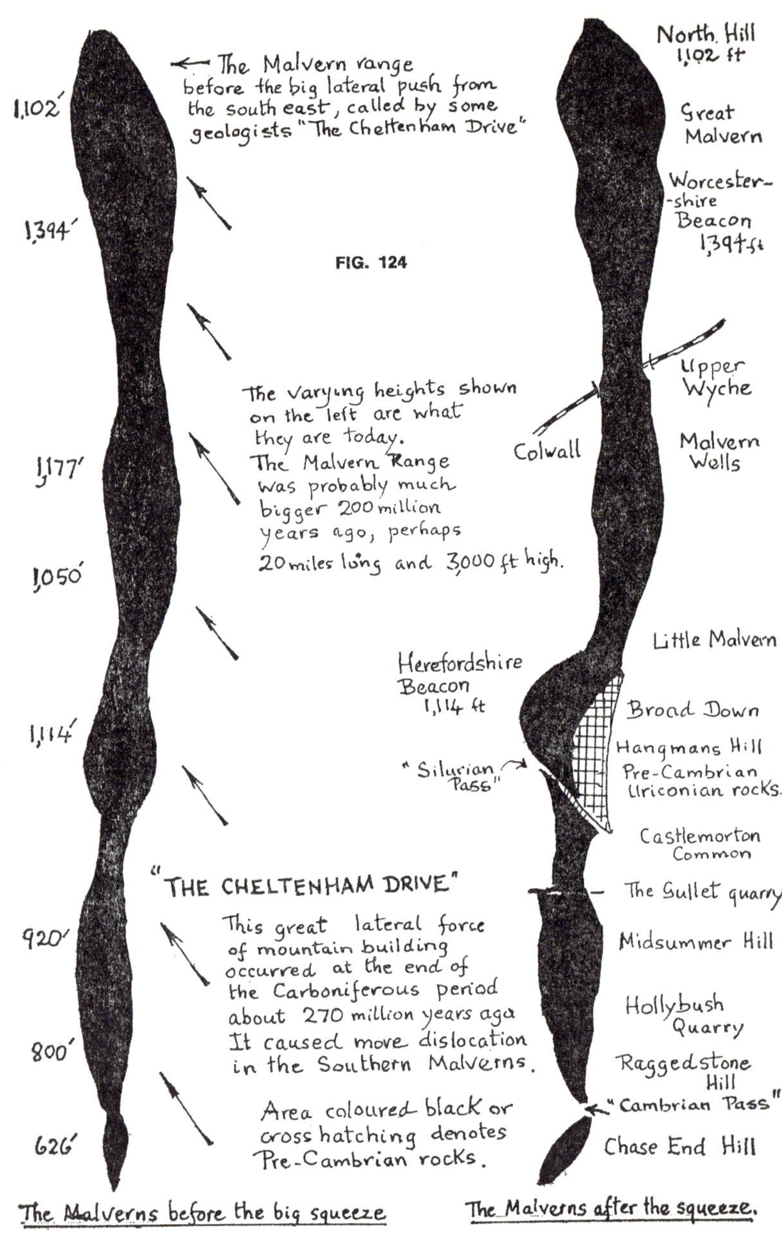

1,102'
1,394'
1,177'
1,050'
1,114'
920'
800'
626'

FIG. 124

← The Malvern range before the big lateral push from the south east, called by some geologists "The Cheltenham Drive"

The varying heights shown on the left are what they are today. The Malvern Range was probably much bigger 200 million years ago, perhaps 20 miles long and 3,000 ft high.

"THE CHELTENHAM DRIVE"

This great lateral force of mountain building occurred at the end of the Carboniferous period about 270 million years ago. It caused more dislocation in the Southern Malverns.

Area coloured black or cross hatching denotes Pre-Cambrian rocks.

The Malverns before the big squeeze

North Hill 1,102 ft
Great Malvern
Worcester-shire Beacon 1,394 ft
Upper Wyche
Colwall
Malvern Wells
Little Malvern
Herefordshire Beacon 1,114 ft
Broad Down
Hangmans Hill Pre-Cambrian Uriconian rocks.
"Silurian Pass" →
Castlemorton Common
The Gullet quarry
Midsummer Hill
Hollybush Quarry
Raggedstone Hill
"Cambrian Pass"
Chase End Hill

The Malverns after the squeeze.

Figure 123 shows how Chase End Hill has been offset from the general north/south alignment but the thrusting has not been as pronounced as at Herefordshire Beacon. Compare this with the plans in Figure 124.

From the summit of Chase End Hill (mainly consisting of hornblende gneiss) it is worth noticing that the main alignment of the Malvern range disappears, and then reappears near the Forest of Dean in the summit of May Hill, which is a resurgence of the 'thrust'.

CAMBRIAN ROCKS OF THE SOUTH MALVERNS

On the western side of the Malverns for about five miles the Silurian rocks outcrop but by the Gullet Quarry and southwards the Cambrian series of rocks come to the surface.

The basal conglomerate is a very hard rock containing pebbles of the Pre-Cambrian and it can be seen in the track leading to the Obelisk from the Gullet Quarry. This is one of the proofs that Malvernian rocks are Pre-Cambrian because this conglomerate represents the beach in the Cambrian sea which washed around the range of mountains which, even at that time, was very ancient.

Next comes the Hollybush sandstone, rather greenish with the colouring of the mineral glauconite, and this can be seen in a small quarry by White Leaved Oak where there is a gap between Ragged Stone Hill and Chase End Hill. As this gap has the Hollybush sandstone outcropping, it could be called the Cambrian Gap to parallel with the Silurian gap further north.

The oblique strips of Silurian sandstone occupying the Silurian gap and those occupying the Cambrian gap are related to the set of oblique NW/SE faults which affected the Malvern range at a much later time than the 'great squeeze'.

THE BRONSIL SHALES

On the western side of the Malverns, by Bronsil Castle, is an area of Cambrian shales which, being softer rocks, form valleys. These shales are divided into the White Leaved Oak shales (which are black) and the Bronsil shales (which are grey).

Figure 124 shows the outcrop of the Cambrian rocks and, as the Hollybush sandstone only outcrops over a small area, it can be taken for granted that most of the area marked Cambrian rock means the presence of shales.

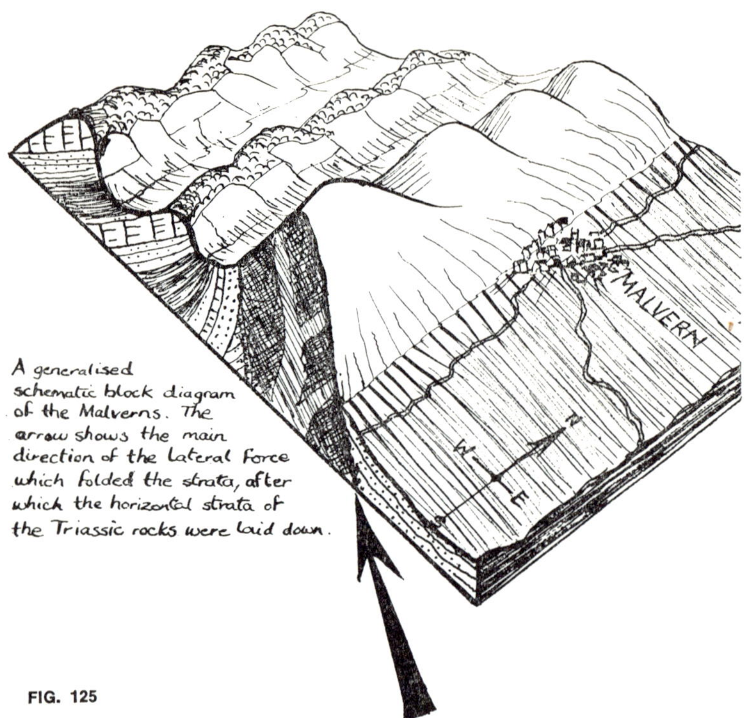

FIG. 125

Figures 126 and 127 show that the shales are dipping very steeply and that magma (molten rock) has been intruded into them. These igneous intrusions form hard outcrops so they can easily be recognised as small knolls crowned with trees. Here again, nature has drawn a convenient geological map. Looking across the White Leaved Oak valley in the shales at least half-a-dozen of these knolls can be seen, some about twenty acres in extent and others merely a few acres.

Figure 128 shows a small igneous outcrop standing out like a castle in the Bronsil area (and previously the site of a medieval castle). The ruins of the castle are now occupied by a prosperous fruit farm, the farmer having the advantage of rich soils derived both from the shales and the igneous rocks—the latter, being mostly basic rocks, are rich in iron.

Figure 128 also shows how the dominant topographical feature

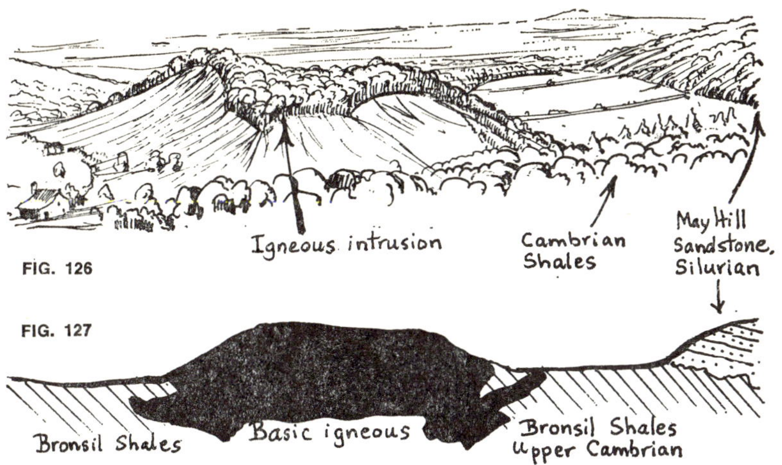

Between Chase End Hill and High Wood.
A wooded Knoll of igneous rock

overshadowing the valley is the plateau formed of the massive May Hill sandstone of the Silurian. In the Bronsil Castle area the plateau slopes down to Eastnor Park, to be followed further westwards by the scarps of the Wenlock limestone and Aymestry limestone.

Is it possible to tell what kind of rocks are underground without going through the very expensive process of boring down into

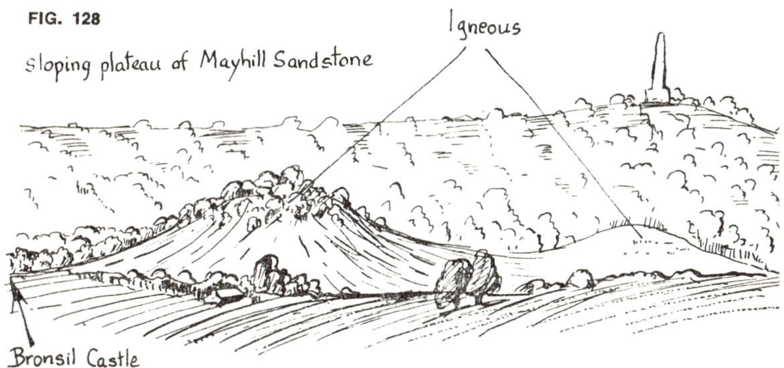

An igneous intrusion in the Bronsil Shales. (Upper Cambrian)
A prosperous fruit farm with abundant water supply is sited on the Bronsil Shales, at Bronsil Castle.

them? The answer is—yes. Structures can be revealed and it is possible to decide whether rocks are light or heavy by what is known as a 'gravity survey'.

Gravity is a force which varies with differences in latitude and variations in topography and the densities of the rocks under the surface. If allowances are made for all these factors then the last one, i.e. the densities of the rocks, can readily be deduced.

A gravity survey of the Malverns area was carried out in 1954. Before it began, careful laboratory checks were made on the densities of the typical rocks of the region and the subsequent field work of the survey covered the plains of Worcestershire and extended on the east as far as the Cotswolds and on the west as far as Wales.

The following were some of the conclusions reached during the survey:

1. The Triassic rocks in the Worcestershire basin are about 5,000 feet thick.
2. There is a pronounced 'step' going down steeply at 45° towards the plain of the Trias.
3. After the gravity-meter observations were plotted on a map they revealed the great scale of the Malvern upheaval.
4. The light rocks lie on the heavier ones and it can be assumed that the 'floor' of the Worcestershire basin underlying the Trias consists of Palaeozoic rocks—but whether these are Silurian or Cambrian it is not possible to decide.

Figure 129 shows two aspects of the results of the gravity survey —the tremendous thickness of the deposits of Trias close to the Malverns and the Pre-Cambrian rock rising like a wall from the plains of the Trias, latter-day evidence of that violent upheaval which occurred so many millions of years ago.

THE ORIGINS OF THE MALVERNS

Ideas about the Malverns are constantly changing as new exposures reveal new clues about their origin. Sometimes these clues are quite dramatic—like those brought to light in the nineteenth century when the railway tunnel was constructed at the Wyche and the cutting revealed new structures. The most recent contribution to the great Malvern debate is the exciting new exposure of the Silurian rocks dipping off the Pre-Cambrian rocks in the Gullet Quarry.

The sequence of diagrams in Figure 130 shows that the ancient Malvern range sank beneath the Silurian sea and the general assumption is that the Silurian rocks once covered the Malverns. Some geologists have found fragments ('clitter') of Mayhill sandstone (Silurian) on top of Herefordshire Beacon and there is other evidence which supports the view put forward in Figure 130.

After the submersion of the Malverns in the Silurian sea there

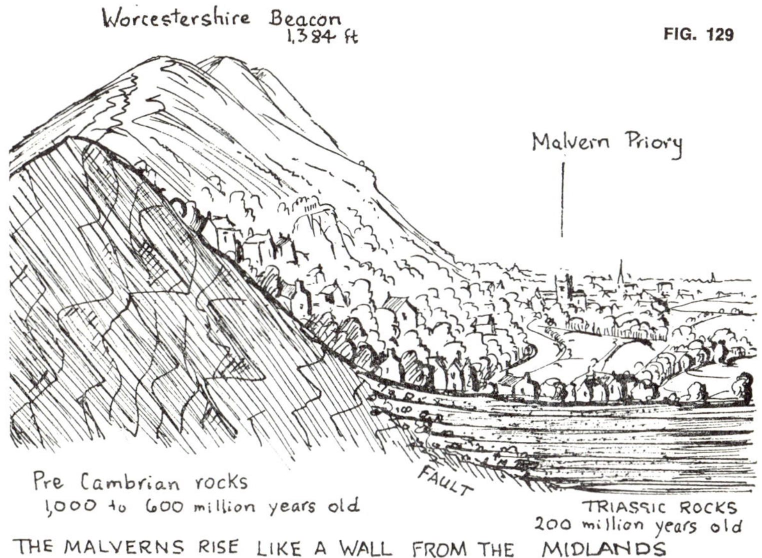

came the 'great squeeze' from the south-east, an upheaval of the rocks which must have occurred even before the laying down of the Triassic rocks because these are not involved in the folds. The Palaeozoic rocks are folded, however, so the folds must have occurred at some time towards the end of the Carboniferous period. In adjacent areas, the later Carboniferous rocks are folded and this is supporting evidence.

It is believed that the 'great squeeze' produced a single monoclinal fold with a steep limb to the west and a flat limb to the east (see Figure 130[b]).

During the squeeze, the mass of Pre-Cambrian rocks forming Herefordshire Beacon was pushed over on to the top of the Silurian strata. During the boring of a well near the British Camp Hotel proof was found that Silurian rocks are underneath the Beacon Hill mass, and along the western edge it has been found that the Silurian rocks close to the Malverns are overturned.

After the folding effect of the great squeeze, the Pre-Cambrian rocks fractured and were upthrust into blocks along a general north/south line of dislocation with oblique faults (the cause of the present-day gaps) setting in later. Figure 130(d) shows how the Malvern range stood out as an island mountain (inselberg) in the vast Triassic desert.

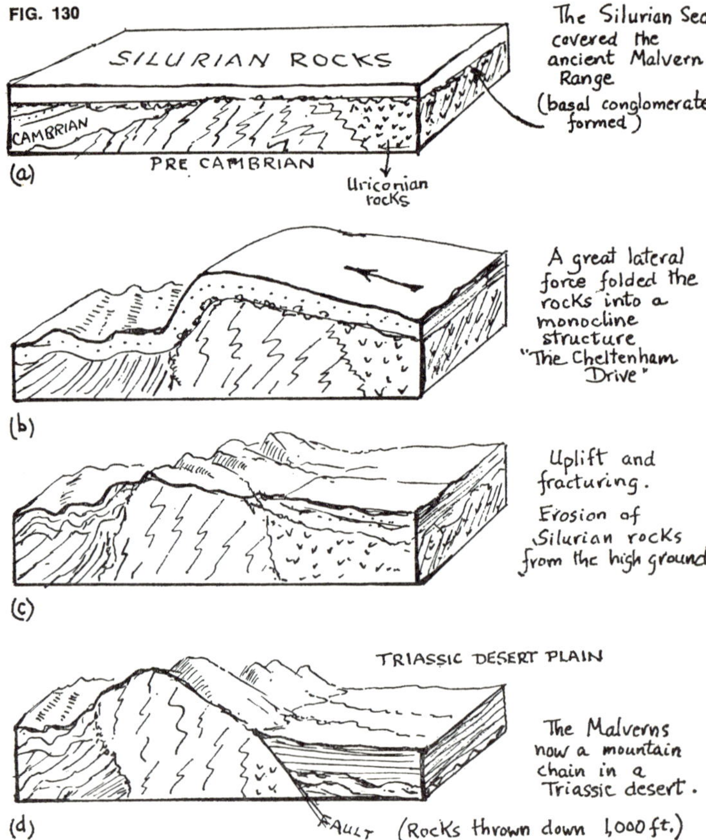

DIAGRAMS SHOWING ONE THEORY AS TO THE ORIGIN OF THE MALVERNS.
(after N. Butcher)

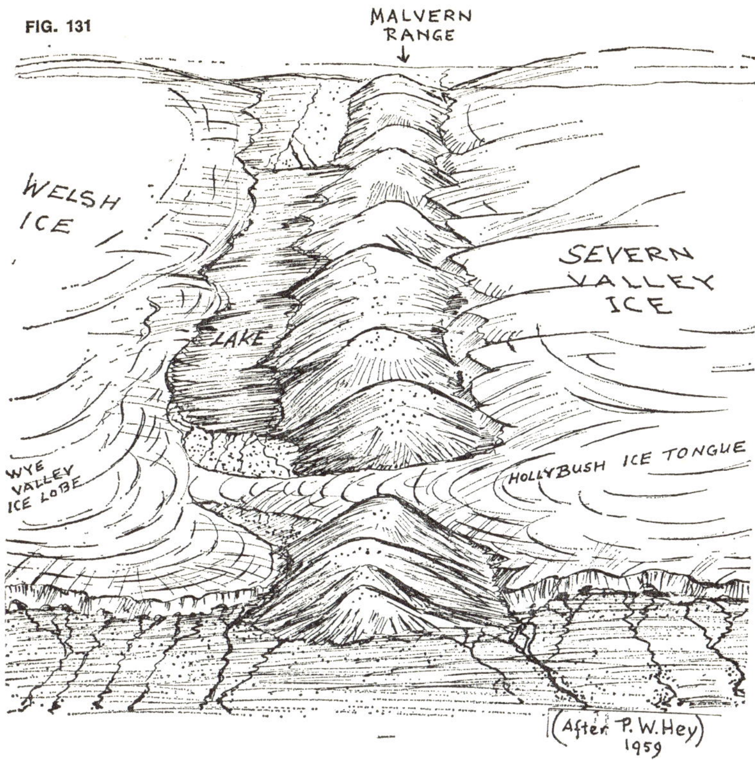

Is there any evidence that the Pre-Cambrian rocks were involved in this folding? This is very difficult to prove for these rocks themselves were folded more than 600 million years ago and were actually part of an ancient Pre-Cambrian range.

A summing-up of current Malvernian theories would be: 'The Malverns are part of a large north/south monoclinal fold involving the Pre-Cambrian and the Palaeozoic rocks and the core of this fold is the Pre-Cambrian rock mass.'

In fact, any geologist coming to Britain for the first time would be well advised to make a visit to the Malverns his very first objective because there is to be seen part of the major structures which form the framework of Britain.

Figure 131, above, depicts the last great phase in the evolution of the Malverns, and is a bird's eye view of the scene, as it was some

M

100,000 years ago during the last but one glacial period of the last Ice Age.

The Severn Valley or Midland ice sheet was fed by ice from the Irish Sea and East Anglia. At the stage shown in Figure 131 the ice was perhaps 300 feet thick. A tongue of ice pushed through the gap by Hollybush Hill and joined up with the Welsh ice sheet. The blockage caused a lake to form on the Western Malverns. Sands and gravels deposited in the lake can be seen in old sandpits at South End, Mathon. The southward drainage became reversed when 'Lake Mathon' was drained. Today we have the Cradley Brook flowing north to join the Severn and Glynch Brook flowing south, thus forming a through valley. Whether the Malverns were covered by an ice sheet is still a point of controversy.

ROCK SPECIMENS

In taking away typical specimens of the rocks of the area, the geologist would select the following:

1. A schist.
2. A granite rock, e.g. a quartz feldspar rock from a pegmatite vein (pegmatite is very coarsely crystallised granite as found in veins).
3. A basic intrusion, e.g. from the Ivy Scar Rock.
4. A piece of massive hornblendic rock.
5. Some of the Palaeozoic rocks as follows:
   a. A piece of May Hill sandstone showing decalcified fossils of brachiopods and corals.
   b. A piece of Cambrian shale.
   c. A piece of Wenlock limestone rich in corals and crinoids.

The geologist might also scare one or two of the local inhabitants by enquiring when the last earthquake took place! The area is the earthquake zone of the South-West Midlands, a fact which is brought to the attention of the inhabitants once or twice during each century when the East Malvern fault decides to settle down just a little more, so causing an earthquake. There has already been one minor earthquake during the early part of this century.

CHAPTER 20

May Hill

This isolated hill dominates the Severn Vale from Tewkesbury to Newnham-on-Severn. It is an outstanding topographical feature which has been made even more conspicuous in recent years by the planting of pine trees. Rising to a height of 971 feet, it is now a National Trust area with a resident warden, and commands such panoramic views that it has become one of Gloucestershire's best-known beauty spots.

GEOLOGICAL BACKGROUND OF MAY HILL

Although May Hill is seemingly cut off from the Malverns it is actually part of the Malvern trend and there are three sets of fold lines in the Palaeozoic rocks—the Malvern trend which is north/

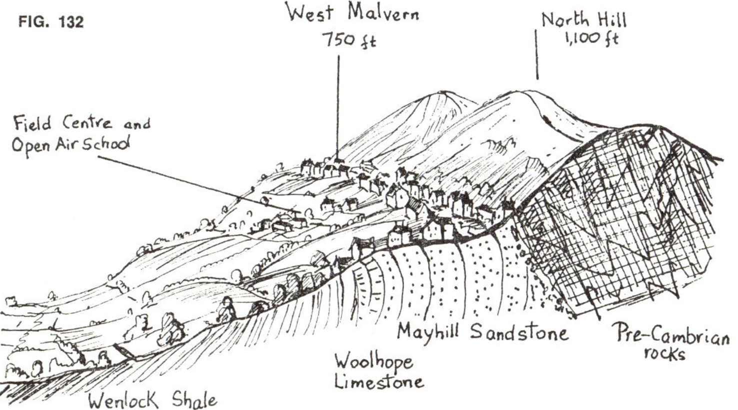

WEST MALVERN   The Silurian May Hill Sandstone forms a high terrace on the West Malverns, but only in the north.

N.B. The sandstones dip vertically and in places are overturned.

south, the Caledonian trend which runs north-east to south-west, and the Armorican trend which runs east-west.

These lines mark general directions of the folds in the rocks and throughout the long history of the Earth (going back over 600 million years) deep-seated earth movements have been occurring along them. The folds are not all exactly parallel—in the area between the Malverns and the Forest of Dean they occur in echelon and often trend to the north/north-west or to the north/north-east.

These are the creases and deformities which have come with the passage of time and have given the face of the country in its old age the form we now recognise and love. The Malverns and May Hill are some of its most striking features and May Hill belongs to one of the folds which forms a dome-shaped structure called a pericline. But as the folds are really lop-sided or asymmetrical, the limb of the western folds is much steeper (often overturned) than the eastern fold.

The Silurian rocks which form the May Hill 'inlier' were at first laid down horizontally, then folded into the dome structure, eroded off to reveal part of the inner core and finally faulted down on both sides. May Hill is an 'inlier' because it is a mass of older rock lying in among newer rock.

This inlier occupies an area of about seven square miles in West

View of May Hill from near Longhope.

The parallel arrangement of wooded limestone scarps appears on the right.

FIG. 133

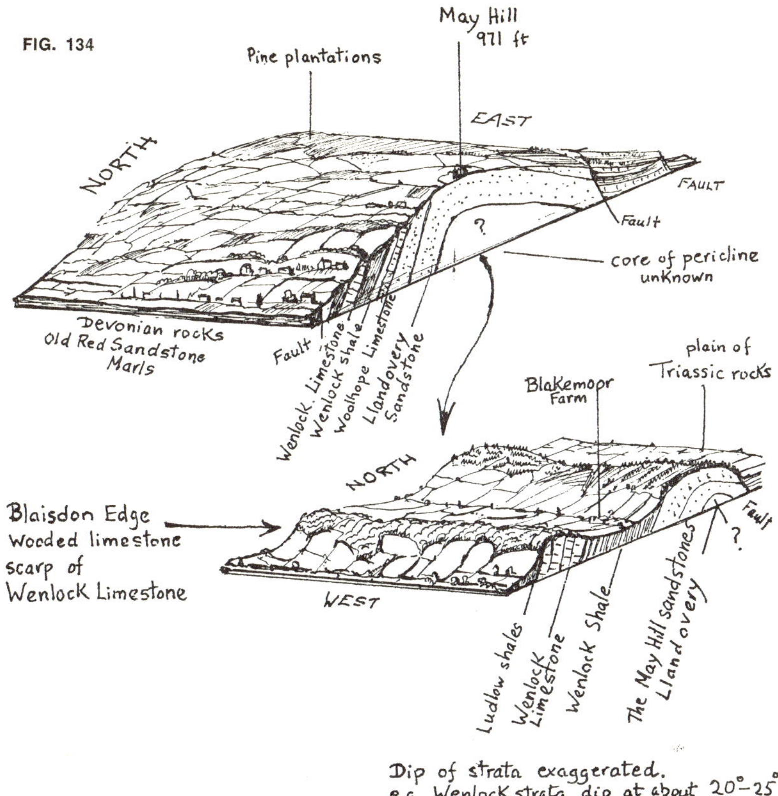

The Silurian rocks of the May Hill pericline.
The second block diagram is a southern continuation of the first one

Gloucestershire and South-East Herefordshire. Although the main view of the hill (see Figure 133) shows it to be merely a rounded whaleback-like shape swelling from the surrounding regions, there are actually very interesting minor variations of relief caused by the great variety of the series of Silurian rocks which outcrop.

The oldest rocks of the Silurian are the Llandovery series forming massive sandstones at least 1,100 feet thick, the same kind of rock which is found on the western flanks of the Malverns by the Obelisk where it is called the May Hill sandstone.

SANDSTONES AND LIMESTONES

On May Hill itself the sandstones vary from conglomerates through coarse-textured rocks right down to fine-textured rock.

Next occurs the Woolhope limestone, about sixty to 200 feet thick, but as it mostly consists of thick argillaceous bands with thin bands of limestones this is not such a distinct scarp-forming feature as the Wenlock limestone.

The Wenlock shales are next in succession. Because these are softer they form the valleys—rather wide valleys because they are between 700 and 800 feet thick.

After the shales occurs the Wenlock limestone, 100 to 350 feet thick, and this forms a distinct scarp, especially in the south of May Hill, where it is known as Blaisdon Edge. These limestones are well exposed in a series of quarries on the eastern side of the Edge known as Hobbs' Quarries, excellent collecting-grounds for corals, crinoids and brachiopods.

Lastly come the Ludlow beds which, being even softer, form the lower ground.

Two block diagrams in Figure 134 show the relationship between these various rock types and the topography of May Hill.

THE VIEW FROM MAY HILL

The view to the west from May Hill comprises a large area of the Forest of Dean where the Carboniferous rocks forming the plateau are surrounded by red rocks of the Devonian series known as the Old Red Sandstone. The scarps rise like a succession of ramparts, and to the north-west the very edge of the coalfield can be seen. This is a superb view for geography teachers wishing to describe to their pupils different regions of Britain. From this hill the following variations of scenery can be observed:

1. The Malvern range—showing it plunging beneath the younger rocks south of Chase Hill.
2. The Forest of Dean coalfield.
3. The Vale of the Lower Severn. The great sweeping bends of the river meanders between Gloucester and Aust can be clearly picked out—glinting ribbons of light on the landscape.
4. The scarp line of the Cotswolds.
5. The plateau of the Welsh mountains with the Brecon Beacons.

Surrounding the hill on the north-east side are very extensive forests of pine, spruce and larch mainly developed on sandstone rocks, but there are many farms and orchards on the west side where numerous springs occur because the Llandovery sandstones which form the main mass of the hill are so porous.

There remains one puzzle—what is the nature of the core of May Hill? Is it like the Malverns but with the Pre-Cambrian rocks lying buried deep down?

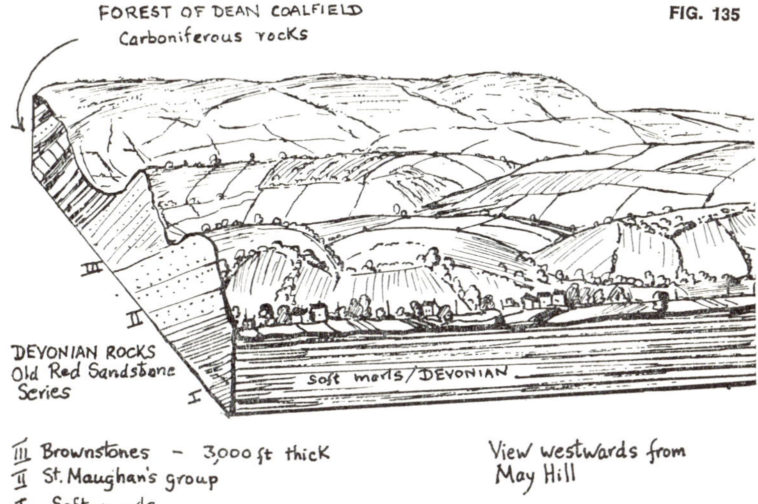

Figure 135 shows the most spectacular view from May Hill, which is to the west and includes a good part of the Forest of Dean plateau. It was actually formed in a basin of Old Red Sandstone and then uplifted to form the present plateau. The basin has two rims formed by two massive bands of the Devonian Sandstone shown in the sketch. The steeply-dipping beds give the appearance of a corrugated rim to the plateau.

WILDERNESS QUARRY

Near Mitcheldean there is a quarry in the Old Red Sandstone which is well worth a visit (see Figure 136).

Although this quarry is not fossiliferous, it shows the red rocks

Wilderness Quarry in Old Red Sandstone, ½m. E. of Mitcheldean
"The Brownstones", Lower O.R.S.

of ancient deserts in a manner which has a drama of its own. Many of the slabs of rock still show sun cracks made 450 million years ago! The seas of the Silurian period which are seen in the rocks of the May Hill area were followed by the great deserts and arid lands which ushered in the Devonian period.

To the geologist, of course, there is no part of the Malverns or May Hill which lacks drama—for there before him are the very first pages of the book of the history of the Earth.

Glossary

BELEMNITE—the hard part of an animal like a sea squid. Belemnites are now extinct and in fossil form are represented by a hard, bullet-like piece of radial calcite called the guard.

BIOTITE—a dark-coloured mica, brown or black, sometimes green; it is abundant in some granites and is also common in schists and gneiss.

BRACHIOPODS—these are small marine invertebrates; there are about 200 species of living brachiopods and about 30,000 fossil forms, brachiopods being among the most abundant Palaeozoic fossils.

CAMBRIAN PERIOD—period of rock formation, 600 million years ago, first period of the Palaeozoic era, named after Wales (Latin, *Cambria*), where rocks of this age were first studied.

CHLORITE—this is one, two, three or more minerals depending on how carefully the constituents are separated; it often forms as an alteration of rocks and can also form in cavities of basic igneous rocks, forming in masses, crusts, fibres or bladed crystals; if considered a single mineral, chlorite is a mixture of magnesium and iron-aluminium silicates, with water.

CLEAVAGE—this is the way some minerals split along planes related to the molecular structure of the mineral and parallel to possible crystal faces.

CRINOIDS—sea 'lilies' (but actually marine animals) which grow in colonies on the sea floor. Some fossil forms were free swimming but most were fixed by a stem.

DIORITE—a basic igneous rock rich in minerals, usually grey or dull green in colour; granites grade into diorites through inter-mediate forms, the granodiorites.

DIP SLOPE—if strata are tilted, the maximum slope is termed the dip slope, at right angles to the strike of the rocks.

ECHINOIDS—sea urchins.

EPIDOTE—this is one of a group of complex silicates of calcium and aluminium with water; it forms in nearly every type of metamorphic rock, in cracks and seams, as crystals or as thin green crusts; it is a typical mineral where igneous rocks have come into contact with limestones.

FELDSPARS—minerals found in nearly all igneous rocks and in rocks formed from them; all are aluminium silicates combined with one or two more metals.

FERRUGINOUS—iron-bearing; ferruginous rocks are those containing iron minerals.

GNEISS—a coarsely-banded metamorphic rock; it can be simply metamorphosed granite or a far more complex rock with possibly four or five different origins, either igneous or sedimentary; it may also include metamorphic rocks which are invaded by igneous materials so that the rock becomes a complex mixture; gneiss is hard to define or describe because it is so varied.

HORNBLENDE—a complex hydrous silicate containing calcium, magnesium and iron and aluminium, which is found in basic igneous rocks but is more often of secondary origin in metamorphosed (i.e. changed) rocks.

IGNEOUS ROCKS—rocks which have been molten at some time in their history, i.e. rocks which all come from magmas, molten mixtures of minerals found deep below the surface of the earth.

METAMORPHIC ROCKS—these are rocks which have been changed; all kinds of rock can be metamorphosed (sedimentary, igneous and other metamorphic rocks); metamorphism results from heat, pressure or permeation by other substances.

MICAS—minerals unusual because of the perfect basal cleavage by which thin flexible sheets can be cleaved off; all include oxides of aluminium and silicon with other metals, singly or in combination.

MIGMATITE—rocks formed by a complex mixture of metamorphic rocks subsequently invaded by igneous rocks.

PALAEOZOIC ERA—the geological era which covers the following geological periods: Permian, Upper Carboniferous, Lower Carboniferous, Devonian, Silurian, Ordovician, Cambrian, in that order of time, the Cambrian rocks being the oldest.

PECTINIDS—a family of bivalve shells, e.g. the common scallop shell.

PRE-CAMBRIAN PERIOD—the vast period of earth history which elapsed before the deposition of the Cambrian fossil-bearing rocks; it covers a period of about 4,000,000,000 years, i.e. approximately 9/10ths of the total age of the earth; this great period of time witnessed the development of the earth, seas and atmosphere, the origin of life, and the early development of living things but Pre-Cambrian animal fossils are rare (it seems likely that Pre-Cambrian animals were soft-bodied and therefore poorly preserved as fossils).

QUARTZ—one of the most common minerals in the earth's crust (Silicon dioxide, SiO_2), an important part of most acid igneous rocks.

SCHISTS—finely-layered metamorphic rocks which split easily; they break in a wavy, uneven surface (this property is called *schistosity*) and they are named after their most characteristic mineral (e.g. mica schist, hornblende schist, chlorite schist, quartz schist).

SEDIMENTARY ROCKS—rocks formed by the accumulation of sediment derived from the breakdown of earlier rocks (e.g. broken down by the action of water or wind), by chemical precipitation or by organic activity; these rocks cover about three-quarters of the earth's surface.

SILURIAN PERIOD—this period of rock formation lasted from 440 to 400 million years ago; it is named after Silures, an ancient tribe of the Welsh borderland, an area where these rocks occur.

SOLIFLUXION—top-soil in tundra climates that moves downhill while the sub-soil is still frozen.

TAELE GRAVELS—gravels not sorted by water action; unstratified gravels formed in glacial periods.

TRIASSIC PERIOD—rocks laid down 230 to 180 million years ago, so named from a three-fold division of its rocks.

TRILOBITES—extinct marine arthropods of great diversity and importance as Palaeozoic guide fossils.

TUFA—redeposited limestone; a calcareous spring will deposit lime over plants on the ground, so forming a petrified mass.

THE GEOLOGICAL AGES OF THE MALVERNS
FIG. 137
Worcestershire Beacon 1,384 ft
West Malvern
Colwall Coppice
Wenlock Limestone 280 ft
Wenlock Shale 640 ft
Woolhope Limestone 50 ft
Mayhill Sandstone 1,100 ft
basal Conglomerate
PRE-CAMBRIAN
420 million years
430 million years
440 million years
older than 600 million years probably 800 to 1,000 million years

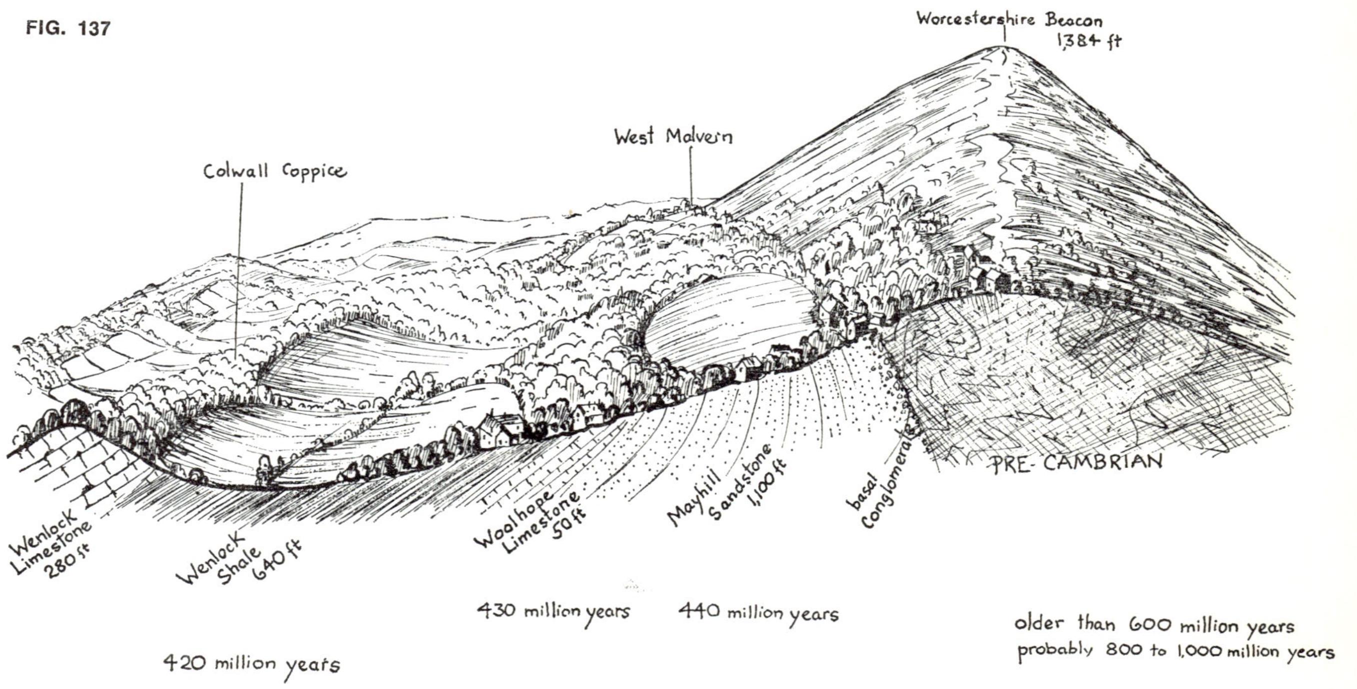

Bibliography

Mesozoic Fossils (British Museum of Natural History): 12s 6d
Palaeozoic Fossils (British Museum of Natural History): 12s 6d
Fossils: A little guide in colour (Paul Hamlyn): 5s
Minerals: A little guide in colour (Paul Hamlyn): 5s
Introducing Geology, D. V. Ager (Faber & Faber): 30s
Geology of the Scenery of England and Wales, Trueman (Pelican
    Books): 5s
Minerals and Rocks, Kirkaldy (Blandford Press): 5s
Bores, Breakers, Waves and Wakes, R. A. R. Tricker (Mills & Boon)

APPENDIX: *Generalised Table of Strata mentioned in this Book*

Era	Period	Millions years ago	Duration	Strata	Main Fossils
Cenozoic	Quaternary Pleistocene	1 to 1½	1 m.y.	Glacial Sands and Gravels, Cheltenham Sands	Palaeolithic implements e.g. flint axes. Bones of mammals and eroded Jurassic fossils. Mammoth bones and teeth.
Mesozoic	Jurassic	The whole Jurassic period began 180 million years ago and ended 135 m.y. ago			Fishes, mammals, reptiles, cycad plants. Ammonites, bivalves, brachiopods.
	Middle Jurassic		Duration 45 m.y.	*Great Oolite series* Great Oolite limestone Taynton Stone Fuller's Earth clay Clypeus Grit Trigonia Grit Notgrove Freestone Gryphite Grit *Inferior Oolite series* Upper Freestone Oolitic Marl Lower Freestone Pea Grit Lower Limestones	*Clypeus ploti* *Trigonia costata* *Gryphaea sublobata* Crinoids, brachiopods, corals
	Lower Jurassic			Cotswold Sands Upper Lias Sands Upper Lias Clay Middle Lias Marlstone rockbed Middle Lias Clays and Sands Lower Lias Clay	Ammonites Jurassic ammonites first arrive
	Triassic	Began 225 m.y. ago	Duration 45 m.y.	*Rhaetic* White Lias Cotham Beds Westbury Shales Bone Bed *Keuper* Tea Green Marl Red Marls Sandstones	Reptiles, fishes, but no ammonites A few plant remains mostly barren of fossils.

	in this area			
	Carboniferous	Ended 270 m.y. ago Began 350 m.y. ago	Coal Measure Shales Sandstones Limestone Red Sandstones and Marls	Plants—forest trees in coal. Corals and brachiopods.
	Devonian	Ended 350 m.y. ago Began 405 m.y. ago		Fish remains
	No Ordovician rocks present in this area			
	Silurian	Began 425 m.y. ago Ended 405 m.y. ago Duration 20 m.y.	Upper Ludlow Shales Aymestry Limestone Lower Ludlow Shales Wenlock Limestone Wenlock Shale Woolhope Limestone Woolhope Shales May Hill Sandstone	First remains of fishes recorded in British rocks. Brachiopods, corals, graptolites.
	Cambrian	Began 600 m.y. ago Ended 500 m.y. ago Duration 100 m.y.	Bronsil Shales White Leaved Oak Shales Hollybush Sandstone Malvern Quartzite	Trilobites
Pre-Cambrian	Not known in the case of the Malvern rocks	Began when the crust of the Earth was formed about 4,500 m. years ago. Malvern rocks are older than 600 m. years and could be up to 1,000 m. years old.	Volcanic rocks known as the Warren House series (of Broad Down and Tinkers Hill). They are similar to the Uriconian volcanic rocks of Shropshire. Malvernian. Crystalline rocks mainly Migmatites.	No recorded fossils. No fossils found in the crystalline rocks of the Malverns. Life began on the Earth about 2,500 million years ago.

NOTES: 1. There are many rocks of the Palaeozoic era not mentioned in the book. These rocks, e.g. the Carboniferous rocks, outcrop at the surface in the Forest of Dean in Gloucestershire, which is mainly a plateau of Carboniferous rocks. The Cretaceous rocks occur to the east of Swindon in Wiltshire, the Chalk country. Ordovician rocks are found in many parts of Wales. Also the Permian rocks are mainly found in N.E. England.

2. The Keuper Marl: this red marl is a misnomer. It is not a marl but, rather, a fine dolomitic silt composed largely of silica stained red with haematite.

The Dolomitic Conglomerate: this is a fossil scree and, in places, the fragments are angular pieces of limestone that should give it the term 'breccia' and not conglomerate.

Trigonia Grit and Gryphite Grit: another misnomer, as it is not a grit but a shelly limestone.

THE SEVERN BORE

F. W. ROWBOTHAM

The how, why, where and when of the spectacular natural phenomenon, illustrated with excellent photographs and including the author's graphic description of riding the bore in a small boat in pitch darkness.

17s 6d

GLOUCESTERSHIRE WOOLLEN MILLS

JENNIFER TANN

A definitive study of the development and decline of the industry. The gazetteer includes an entry for every known workshop and dyehouse and includes a brief survey of what remains at each site. Illustrated with half-tones and maps.

45s

DAVID & CHARLES : NEWTON ABBOT